Medicinal Properties and Molecular Mechanisms of Thai Traditional Herbs

With a focus on the unique cultural and ecological aspects of Thailand, this latest title in the popular series, *Natural Products Chemistry of Global Plants*, offers invaluable insights for researchers and healthcare professionals interested in exploring the potential of Thai herbal medicines in the broader context of global health and wellness. It also appeals to individuals seeking natural and traditional alternatives to modern medicine. The book provides a bridge between modern pharmacology and traditional medicinal practices, thereby bringing Thai traditional herbal medicines into the mainstream, while giving a critical explanation of their efficacy and safety.

Natural Products Chemistry of Global Plants

Series Editor: Clara Bik-san Lau

Founding Editor: Raymond Cooper

This unique book series focuses on the natural products chemistry of botanical medicines from different countries such as Bangladesh, Borneo, Brazil, Cameroon, China, Ecuador, India, Iran, Laos, Romania, Sri Lanka, Turkey, etc. These fascinating volumes are written by experts from their respective countries. The series will focus on the pharmacognosy, covering recognized areas rich in folklore as well as botanical medicinal uses as a platform to present the natural products and organic chemistry. Where possible, the authors will link these molecules to pharmacological modes of action, reflecting the ethnopharmacological uses. The series intends to trace a route through history from ancient civilizations to the modern day showing the importance to man of natural products in medicines, in foods and a variety of other ways. With special emphasis on plant parts for medicinal uses, phytochemistry and biological activities, this book series will be of useful reference to scientists/ pharmacognosists/ pharmacists/ chemists/ graduates/ undergraduates/ researchers in the fields of natural products, herbal medicines, ethnobotany, pharmacology, chemistry and biology, Furthermore, pharmaceutical companies may also found valuable information on potential herbs and lead compounds for the future development of health supplements and western medicines.

Recent Titles in this Series:

Natural Products Chemistry of Botanical Medicines from Cameroonian Plants, *Xavier Siwe Noundou*

Medicinal Plants and Mushrooms of Yunnan Province of China, *Clara Lau and Chun-Lin Long*

Medicinal Plants of Borneo, *Simon Gibbons and Stephen P. Teo*

Natural Products and Botanical Medicines of Iran, *Reza Eddin Owfi*

Natural Products of Silk Road Plants, *Raymond Cooper and Jeffrey John Deakin*

Brazilian Medicinal Plants, *Luzia Modolo and Mary Ann Foglio*

Medicinal Plants of Bangladesh and West Bengal: Botany, Natural Products, and Ethnopharmacology, *Christophe Wiart*

Traditional Herbal Remedies of Sri Lanka, *Viduranga Y. Waisundara*

Medicinal Plants of Ecuador, *Pablo A. Chong Aguirre, Patricia Manzano Santana, Migdalia Miranda Martínez (Eds)*

Medicinal Plants of Laos, *Djaja Djendoel Soejarto, Bethany Gwen Elkington and Kongmany Sydara*

Edible and Medicinal Mushrooms of the Himalayas: Climate Change, Critically Endangered Species and the Call for Sustainable Development, *Ajay Sharma, Garima Bhardwaj and Gulzar Ahmad Nayik*

Medicinal Plants of Turkey, *Ufuk Koca-Caliskan and Esra Akkol*

Natural Products and Medicinal Properties of Carpathian (Romanian) Plants, *Adina-Elena Segneanu*

Medicinal Plants of Sabah, North Borneo, *Christophe Wiart*

Medicinal Properties and Molecular Mechanisms of Thai Traditional Herbs, *James M. Brimson, Tewin Tencomnao, and Ciro Isidoro*

Medicinal Properties and Molecular Mechanisms of Thai Traditional Herbs

Edited by James M. Brimson, Tewin Tencomnao, and Ciro Isidoro

CRC Press
Taylor & Francis Group
Boca Raton London New York

CRC Press is an imprint of the
Taylor & Francis Group, an **informa** business

Designed cover credit: James M. Brimson

First edition published 2025
by CRC Press
2385 Executive Center Drive, Suite 320, Boca Raton, FL 33431

and by CRC Press
4 Park Square, Milton Park, Abingdon, Oxon, OX14 4RN

CRC Press is an imprint of Taylor & Francis Group, LLC

ISBN: 978-1-032-82977-7 (hbk)
ISBN: 978-1-032-82980-7 (pbk)
ISBN: 978-1-003-50872-4 (ebk)

DOI: 10.1201/9781003508724

Typeset in Times
by Deanta Global Publishing Services, Chennai, India

Contents

Editors vii
Contributors x
Preface xiii
Acknowledgements xiv

Chapter 1 A Brief History of Thai Traditional and Herbal Medicine 1

James M. Brimson, Sirikalaya Brimson, Premrutai Thitilertdecha, Tewin Tencomano, and Ciro Isidoro

Chapter 2 Methods for Extracting Medicinal Compounds from Thai Herbs 11

James M. Brimson, Sirikalaya Brimson, and Tewin Tencomnao

Chapter 3 Thai Medicinal Uses of Agarwood (*Aquilaria crassna*) 23

Apinan Senabunyarith, Phaniendra Alugoju, Madugula Shabareesh, Siriporn Chuchawankul, James M. Brimson, and Tewin Tencomnao

Chapter 4 *Rhinacanthus nasutus* (L.) Kurz's Role in Cancer and Neurodegenerative Disease: Prevention and Treatment 34

James M. Brimson and Tewin Tencomano

Chapter 5 The Role of *Clerodendrum petasites* S. Moore in Disease Treatment 44

Premrutai Thitilertdecha, James M. Brimson, and Nattawat Onlamoon

Chapter 6 The Medicinal Properties of Resveratrol, a Polyphenol Component of the Thai Silkworms' Plant *Morus alba* 54

Alessandra Ferraresi, Chiara Vidoni, and Ciro Isidoro

Chapter 7 Thai Traditional and Medicinal Uses of *Zingiber officinale* Roscoe 72

Shu-qin Cao, James M. Brimson, and Tewin Tencomano

Chapter 8 *Punica granatum* L. in Thai Traditional Medicine 83

Madhura B. de Silva, James M. Brimson, and Tewin Tencomnao

Chapter 9 The Memory-Enhancing Effects of Thai Medicinal Herb *Bacopa monnieri* 93

James M. Brimson

Chapter 10 Traditional Thai Medicinal Uses of *Curcuma comosa* Roxb. 100

Madhura B. de Silva, James M. Brimson, and Tewin Tencomnao

Chapter 11 Traditional Thai Medicinal Properties of *Cleistocalyx nervosum* var. paniala 106

Rajasekharan Sharika, Mani Iyer Prasanth, Bhagavathi Sundaram Sivamaruthi, and Tewin Tencomnao

Chapter 12 Kratom: A Beneficial Traditional Thai Medicine or a Harmful Narcotic Drug? 114

Clerance Su Yee Cheong and Tewin Tencomnao

Chapter 13 *Piper nigrum L* in Thai Traditional Medicine 139

Mani Iyer Prasanth, Kanika Verma, Rajasekharan Sharika, Pankaj Nair Gayathri, Infanta Jereena Leonard, Tewin Tencomnao, and Siriporn Chuchawankul

Chapter 14 Medicinal Roles and Health Benefits of *Citrus hystrix* DC. (Kaffir lime) 160

Anchalee Prasansuklab, Kamonwan Chaikhong, and Sawarin Chumpolphant

Chapter 15 Updated Medicinal Uses of Marijuana in Thailand (*Cannabis* species in Thai Traditional Medicine) 182

Mani Iyer Prasanth, Bhagavathi Sundaram Sivamaruthi, Rajasekharan Sharika, Infanta Jereena Leonard, Pankaj Nair Gayathri, Tewin Tencomnao, and Siriporn Chuchawankul

Index 205

Editors

James M. Brimson is a researcher and lecturer in the Faculty of Allied Health Sciences at Chulalongkorn University, Bangkok, Thailand. James started his research career while an undergraduate at the University of the West of England (U.W.E.), where he visited the lab of Prof. Gregorio Gil in 2004 for a one-year research internship at Virginia Commonwealth University (V.C.U.) in the United States of America, before returning to finish his B.Sc. in Applied Genetics at U.W.E. He earned his Ph.D. in 2010 from the University of Bath, where he specialised in the pharmacology of the sigma-1 receptor in the Department of Pharmacy and Pharmacology, under the guidance of Prof. Stephen Safrany. His thesis, titled 'The Pharmacology of the Sigma-1 Receptor,' laid the foundation for his extensive research in this field.

Following his doctoral studies, Dr. Brimson pursued postdoctoral research in the laboratory of Professor Tewin Tencomnao, where he deepened his investigation into sigma receptors. During this time, he also expanded his research focus to include natural products, exploring their potential therapeutic applications in neurological disorders. His work has contributed to a greater understanding of the sigma-1 receptor's role in neuroprotection, cognitive enhancement, and its interaction with various natural compounds.

Dr. Brimson's research is interdisciplinary, bridging the gap between pharmacology, neuroscience, and natural product chemistry. He is particularly interested in the therapeutic potential of natural compounds in treating neurodegenerative diseases, with a focus on the mechanisms of action involving sigma receptors. At Chulalongkorn University, he continues to mentor students and collaborate with international researchers, advancing the field of sigma receptor research and its applications in modern medicine. Dr. Brimson was one of only five people to be awarded the Mid-Career Scholarship Fund from the National Research Council of Thailand in 2024. Dr. Brimson has published over 50 manuscripts, has a H-Index of 20, with over 1000 citations, and continues to be very active with his research activities.

In addition to his academic pursuits, Dr. Brimson is an active member of the scientific community, frequently presenting his research at international conferences and contributing to peer-reviewed journals. Dr. Brimson is an editor for *Geriatric Pharmacology* and has served as associate editor for *Journal of Alzheimer's Disease*. His work has been recognised for its impact on understanding the complex roles of sigma receptors in health and disease, and he remains dedicated to uncovering new therapeutic strategies for neurological conditions.

Ciro Isidoro is a distinguished expert in the field of pathology, serving as a Professor of General Pathology and the head of the Laboratory of Molecular Pathology and Nanobioimaging at the Department of Health Sciences, Università del Piemonte Orientale in Novara, Italy. He holds a Doctor of Science (D.Sc.) degree in Biological Sciences with Summa Cum Laude from the Università di Torino and a Doctor of

Medicine and Surgery (M.D.) degree with Summa Cum Laude from the Università del Piemonte Orientale. His academic journey includes a Ph.D. fellowship as an Assistant Researcher at the Institut für Pathobiochemie, Westfälische Wilhelms-Universität Münster in Germany, where he developed a strong foundation in molecular pathology.

Professor Isidoro's contributions to the scientific community have been recognised internationally, including his appointment as Professor Honoris Causa by the Faculty of Medicine and Pharmacy at the Université de la Franche-Comté in Besançon, France. He also holds prestigious visiting professorships at the Siriraj Faculty of Medicine, Mahidol University in Bangkok, Thailand, and at the Department of Cell Biology, Oklahoma University Health Science Center in Oklahoma City, USA.

An active participant in the global scientific community, Professor Isidoro serves on the editorial boards of numerous high-impact journals, including *Autophagy*, *Molecular Carcinogenesis*, *Cancers*, *Frontiers in Oncology*, *Cells*, *Biomedicines*, *International Journal of Molecular Sciences*, *Discover Oncology*, among others. He is also the Editor-in-Chief of *Journal of Traditional and Complementary Medicine* and *Journal of Cancer and Metastasis Treatment*. His editorial work reflects his commitment to advancing the field and ensuring the dissemination of cutting-edge research.

Professor Ciro Isidoro is a globally recognised expert in autophagy and lysosome biogenesis. Professor Isidoro's research focuses on the role of the autophagy-lysosomal degradation pathway in cancer and neurodegenerative diseases, as well as the involvement of the endocytic and exocytic pathways in the tumour microenvironment and neuronal plasticity. His laboratory also studies the epigenetic mechanisms of natural substances with anti-tumour nutraceutical properties. His work has provided critical insights into the cellular mechanisms of autophagy underlying cancer cell dormancy and metastasisation, with implications for developing novel therapeutic strategies. His interdisciplinary approach combines molecular pathology with advanced imaging techniques, contributing to a deeper understanding of the cellular processes that govern health and disease. With over 200 peer-reviewed publications, an H-Index of 50 and more than 24.000 citations (Google Scholar), Professor Isidoro was ranked among the world's top 2% scientists in 2023 and recognised as a leading Italian scientist in biomedical sciences.

Through his leadership, teaching, and research, Professor Isidoro has made significant contributions to the field of molecular pathology, mentoring the next generation of scientists and fostering international collaborations that drive scientific innovation.

Tewin Tencomnao is a prominent researcher and assistant professor in the Department of Clinical Chemistry, and Dean of the Faculty of Allied Health Sciences, Chulalongkorn University, Bangkok, Thailand. He earned his Ph.D. in Biochemistry and Molecular Biophysics from Virginia Commonwealth University, School of Medicine, Medical College of Virginia, in the United States of America. His research career is marked by a focus on the molecular mechanisms by which natural products combat neurodegeneration and ageing, employing various study models, including cell cultures and the *in vivo* model organism *C. elegans*.

Dr. Tencomnao's work has garnered significant recognition, including the prestigious DMSc Award in 2015 from Thailand for his research on the 'Effect and Underlying Mechanisms of *Rhinacanthus nasutus* against Neurodegeneration.' Additionally, he was honoured with the Gold Award in 2015 at the International Engineering Invention & Innovation Exhibition (i-ENVEX) in Malaysia for his development of an anti-ageing cream utilizing natural resveratrol extracted from the Perlis grapevine. In 2001 Dr. Tencomnao won the prestigious Association of Universities of Asia and Pacific (AUAP) Research Fund Award for the Evaluation of the neuroprotective effects of Mitragynaspeciosa extract using a *C. elegans* model.

In his laboratory, Dr. Tencomnao leads groundbreaking research that bridges the gap between traditional medicine and modern molecular biology, focusing on identifying and understanding the therapeutic potential of natural compounds in treating neurodegenerative diseases and ageing-related conditions. His innovative approach and commitment to research excellence have established him as a leader in his field, with his work contributing to the development of novel therapeutic strategies and enhancing the scientific understanding of natural product pharmacology.

Contributors

Phaniendra Alugoju
Center of Excellence on Natural Products for Neuroprotection and Anti-Ageing
Chulalongkorn University
Bangkok, Thailand
and
Department of Clinical Chemistry
Faculty of Allied Health Sciences
Chulalongkorn University
Bangkok, Thailand

James M. Brimson
Research, Innovation and International Affairs
Faculty of Allied Health Sciences
Chulalongkorn University
Bangkok, Thailand
and
Center of Excellence on Natural Products for Neuroprotection and Anti-Ageing
Chulalongkorn University
Bangkok, Thailand

Sirikalaya Brimson
Department of Clinical Microscopy
Faculty of Allied Health Science
Chulalongkorn University
Bangkok, Thailand

Shuqin Cao
Department of Clinical Chemistry
Faculty of Allied Health Sciences
Chulalongkorn University
Bangkok, Thailand

Kamonwan Chaikhong
Interdisciplinary Graduate Program in Biomedical Science
Graduate School
Chulalongkorn University
Bangkok, Thailand
and
Center of Excellence on Natural Products for Neuroprotection and Anti-Ageing
Chulalongkorn University
Bangkok, Thailand

Clerance Su Yee Cheong
Center of Excellence on Natural Products for Neuroprotection and Anti-Ageing
Chulalongkorn University
Bangkok, Thailand
and
Department of Clinical Chemistry
Faculty of Allied Health Sciences
Chulalongkorn University
Bangkok, Thailand

Sawarin Chumpolphant
Graduate Program in Clinical Chemistry and Molecular Medicine
Faculty of Allied Health Science
Chulalongkorn University
Bangkok, Thailand
and
Center of Excellence on Natural Products for Neuroprotection and Anti-Ageing
Chulalongkorn University
Bangkok, Thailand

Siriporn Chuchawankul
Department of Transfusion Medicine and Clinical Microbiology
Faculty of Allied Health Sciences
Chulalongkorn University
Bangkok, Thailand
and
Immunomodulation by Natural Products Research Unit
Chulalongkorn University
Bangkok, Thailand

Madhura B. de Silva
Center of Excellence on Natural Products for Neuroprotection and Anti-Ageing
Chulalongkorn University
Bangkok, Thailand

Alessandra Ferraresi
Laboratory of Molecular Pathology
Department of Health Sciences
Università del
and
Piemonte Orientale
Novara, Italy

Pankaj Nair Gayathri
Department of Biotechnology
Nehru Arts and Science College
Coimbatore, India

Ciro Isidoro
Laboratory of Molecular Pathology
Department of Health Sciences
Università del Piemonte Orientale
Novara, Italy

Infanta Jereena Leonard
Department of Biotechnology
Nehru Arts and Science College
Coimbatore, India

Nattawat Onlamoon
Siriraj Research Group in Immunobiology and Therapeutic Sciences
Faculty of Medicine Siriraj Hospital
Mahidol University
Bangkok, Thailand

Anchalee Prasansuklab
College of Public Health Sciences
Chulalongkorn University
Bangkok, Thailand
and
Center of Excellence on Natural Products for Neuroprotection and Anti-Ageing
Chulalongkorn University
Bangkok, Thailand

Mani Iyer Prasanth
Department of Clinical Chemistry
Faculty of Allied Health Sciences
Chulalongkorn University
Bangkok, Thailand
and
Center of Excellence on Natural Products for Neuroprotection and Anti-Ageing
Chulalongkorn University
Bangkok, Thailand

Apinan Senabunyarith
Center of Excellence on Natural Products for Neuroprotection and Anti-Ageing
Chulalongkorn University
Bangkok, Thailand
and
Department of Clinical Chemistry
Faculty of Allied Health Sciences
Chulalongkorn University
Bangkok, Thailand

Madugula Shabareesh
Department of Biotechnology
Sri Venkateswara Institute of Medical Sciences
Tirupathi, India

Rajasekharan Sharika
Immunomodulation of Natural Products Research Unit
Faculty of Allied Health Sciences
Chulalongkorn University
Bangkok, Thailand
and
Department of Transfusion Medicine and Clinical Microbiology
Faculty of Allied Health Sciences
Chulalongkorn University
Bangkok, Thailand

Bhagavathi Sundaram Sivamaruthi
Innovation Center for Holistic Health, Nutraceuticals and Cosmeceuticals
Faculty of Pharmacy
Chiang Mai University
Chiang Mai, Thailand

Tewin Tencomano
Center of Excellence on Natural Products for Neuroprotection and Anti-Ageing
Chulalongkorn University
Bangkok, Thailand
and
Department of Clinical Chemistry
Faculty of Allied Health Sciences
Chulalongkorn University
Bangkok, Thailand

Premrutai Thitilertdecha
Siriraj Research Group in Immunobiology and Therapeutic Sciences
Faculty of Medicine Siriraj Hospital
Mahidol University
Bangkok, Thailand

Kanika Verma
Natural Products for Neuroprotection and Anti-Ageing Research Unit
Chulalongkorn University
Bangkok, Thailand
and
Department of Clinical Chemistry
Faculty of Allied Health Sciences
Chulalongkorn University
Bangkok, Thailand

Chiara Vidoni
Laboratory of Molecular Pathology
Department of Health Sciences
Università del Piemonte Orientale
Novara, Italy

Preface

Thai traditional herbal medicine has a history spanning centuries, intricately woven into Thailand's cultural and spiritual fabric. The knowledge passed down through generations offers a unique blend of wisdom, observation, and empirical practice, which modern science continues to explore with increasing interest. In recent years, the global surge in research on natural products has placed Thai herbs under the spotlight for their potential therapeutic benefits and molecular mechanisms of action. This book, *Medicinal Properties and Molecular Mechanisms of Thai Traditional Herbs,* is an interdisciplinary effort to bridge the rich heritage of Thai traditional medicine with the rigor of contemporary biomedical science. The work is a collaborative endeavour, bringing together experts in pharmacology, biochemistry, molecular biology, human pathology, and traditional medicine to highlight several key Thai herbs' medicinal properties and molecular mechanisms. The chapters provide an in-depth exploration of each herb's traditional uses, extraction techniques, bioactive compounds, and their mechanisms of action in the context of modern pharmacological understanding. Starting with a historical overview in Chapter 1, the book traces the roots of Thai traditional and herbal medicine, setting the stage for a scientific discourse. From the role of *Agarwood (Aquilaria crassna)* in Chapter 3 to the anti-cancer and neuroprotective properties of *Rhinacanthus nasutus* in Chapter 4, the book delves into various plants and their applications. The unique contributions of *Clerodendrum petasites*, *Zingiber officinale*, and *Curcuma comosa* are discussed alongside the well-known therapeutic effects of compounds like resveratrol from *Morus alba* (Chapter 6) and the memory-enhancing *Bacopa monnieri* (Chapter 9). The latter chapters expand the narrative to include contemporary and sometimes controversial topics, such as the medicinal properties of kratom (*Mitragyna speciosa*, Chapter 12) and the evolving legal and medicinal landscape surrounding marijuana (*Cannabis* species, Chapter 15). Each chapter underscores the importance of understanding these herbs not only through the lens of tradition but also through modern pharmacological research, which validates and broadens their potential applications. This book serves as a resource for researchers, healthcare professionals, and anyone interested in natural product medicine. It highlights the profound connection between traditional knowledge and scientific innovation, encouraging a holistic approach to understanding and harnessing the medicinal potential of Thai herbs. We, the editors, are deeply grateful to the contributors for their expertise and dedication and to the many researchers and practitioners who have paved the way for this field of study. We hope this book inspires further research and fosters an appreciation of the wealth of medicinal knowledge inherent in traditional Thai herbal medicine.

Acknowledgements

The editors would like to thank all the people who contributed to this book, particularly the photographers who kindly provided the images: Mr. Supakorn Yuenyongwannachot, Dr. Onusa Thamsermang, Mr. James Phungphol, Assoc. Prof. Dr. Theerakamol Pengsakul, and Mr. Wudtipong Vongthip.

We thank www.Biorender.com and www.Chemaxon.com for the scientific image/chemical structure drawing software. We thank www.canva.com for the image drawing and design software.

In memory of Professor Nijsiri Ruangrungsi.

1 A Brief History of Thai Traditional and Herbal Medicine

James M. Brimson, Sirikalaya Brimson, Premrutai Thitilertdecha, Tewin Tencomano, and Ciro Isidoro

1.1 INTRODUCTION

A system of traditional medicine in Thailand (Figure 1.1) has existed since the Ayutthaya period (1350–1767 A.D.) (Saralamp et al. 1996). In the year 1888, during the reign of King Rama V, Siriraj Hospital was established with the intent to combine modern medicine with the traditional medicine system. By the year 1913, the traditional medicine system was discontinued because of the contrast among different doctrines. However, with time, botanical medicine gained increasing importance in medical treatments, and by 1999 the first list of essential medicines derived from processed botanical products, in the form of decoctions, powders, or extracts from leaves, roots, or bark, was approved.

In 2002, the Ministry of Public Health was established in Thailand, and herbal medicine, along with other practices of traditional medicine, was officially included in the medical system. In 2012, 52 traditional drugs and 21 single herbal drugs were approved for use in Thai traditional medicine. However, concerns have been raised about the quality control in the preparation of herbal extracts and their safe use. Possible adverse side effects may be linked to the processing of the herbal source to produce the extract or the powder. Another important issue is related to the authentication of the herbal source; herbs of different species, though belonging to the same genus, have different biochemical compositions and may show different efficacy and toxicity. The cultivation soil of the medicinal plant also influences its final composition and therefore its curative efficacy and possible toxicity. Thus, to guarantee efficacy and safety in medical treatment it is of paramount importance to establish standards for the source, processing, and characterisation of natural products.

Currently, silymarin is the only herbal drug that has been licenced as a "modern herbal medicine" based on its safety, efficacy, and quality.

Most herbal products cannot be considered and prescribed as medicine because of lack of evidence of their efficacy as proved in clinical trials, and because of unpredictable adverse effects arising from their processing.

DOI: 10.1201/9781003508724-1

FIGURE 1.1 Map of Thailand and a selection of herbs grown and consumed in Thailand for their health benefits.

Below (in Table 1.1) we report some Thai plants that are widely used in the kitchen for the preparation of food, and that are well known for their beneficial health effects.

1.2 TRADITIONAL MEDICINAL USE OF THAI HERBS

1.2.1 *Centella asiatica*

Stems and leaves of *Centella asiatica* (Figure 1.2) are boiled and juice extracted, which is then drunk to reduce "internal heat," aid in the healing of internal trauma, reduce redness and swelling from inflammation, reduce fever, relieve headaches from migraine, sooth stomach ache, alleviate diarrhoea, and alleviate jaundice from hepa titis. They also have physiological actions such as wound healing and diuretic effects. *C. asiatica* can be mashed up and used as a topical medicine for various skin conditions such as leprosy, eczema, abscesses, and measles rash. They potentially remove or dissolve bladder stones (Termwiset et al. 2013; Gohil, Patel, and Gajjar 2010).

1.2.2 *Rhinacanthus nasutus*

Rhinacanthus nasutus (Figure 1.3) roots are boiled and eaten to reduce fever, cause diuretic effects, reduce skin rash from inflammation, and alleviate leprosy. *R. nasutus* has mild antibiotic effects on tuberculosis and antifungal effects, especially in skin conditions caused by fungal infections such as ringworm. Boiled-extracted water from roots helps reduce dandruff or dry scalp from fungal infections. Fresh leaves and roots are ground up, soaked in alcohol, and used as topical medicine to relieve eczema, ringworm, chloasma or melanoderma (brown spots from hormonal changes), and tinea. Cooked or boiled *R. nasutus* leaves are consumed to relieve haemorrhoids and constipation and eliminate intestinal parasites (Termwiset et al. 2013; Shamnan 2013).

TABLE 1.1
Thai Plants Widely Used for Cooking in the Kitchen

Thai Plants	Beneficial Effect	References
Turmeric (*Curcuma longa* L.)	Digestive disorder Arthritis Anti-cancer (e.g. anti-proliferation, anti-metastasis, angiogenesis, and induct apoptosis) Skin disease Delays diabetic cataract Anti-oxidant properties Autoimmune disease Anti-inflammatory effect Neuroprotective effect	Reviewed in Pari, Tewas, and Eckel 2008
Asiatic pennywort (*Centellaasiatica* (L.) Urb.)	Anti-tumour Anti-oxidant properties Wound healing Alzheimer's disease	Babu, Kuttan, and Padikkala 1995; Veerendra Kumar and Gupta 2003; Parameshwaraiah and Shivakumar 1998; Gnanapragasam et al. 2004
Ginger (*Zingiberofficinale* Roscoe)	Antioxidant Anti-cancer Balance the immune system Cardiovascular protection Digestive health	Butt and Sultan 2011
Ivy gourd (*Cocciniagrandis* L.)	Blood sugar-lowering effect Antimicrobial activity Antidyslipidemic activity Source of natural antioxidant	Singh et al. 2007; Umamaheswari and Chatterjee 2008; Satheesh and Murugan 2011; Munasinghe et al. 2011
Shallot (*Allium ascalonicum* L.)	Anti-oxidative and free radical scavenging abilities Against cyclosporine A nephrotoxicity Against human leukemia cells	Ogra et al. 2005; Wongmekiat, Leelarugrayub, and Thamprasert 2008
Common basil (*Ocimumbasilicum* L.)	Prevention of cardiovascular disease Antioxidant properties Prevention of stroke Hepatoprotective activity Anti-inflammatory effects Antimicrobial activities	Marzouk 2009; Benedec et al. 2007; Bora, Arora, and Shri 2011; Kaurinovic et al. 2011; Fathiazad et al. 2012; Kaya, Yigit, and Benli 2008
Kitchen mint (*Menthacordifolia*Opiz ex Fresen)	Itchy skin	
Leech lime (*Citrus hystrix* DC.)	Inhibitors of nitric oxide generation Antifertility effect	Piyachaturawat, Glinsukon, and Chanjarunee 1985; Murakami et al. 1999

(Continued)

TABLE 1.1 CONTINUED
Thai Plants Widely Used for Cooking in the Kitchen

Thai Plants	Beneficial Effect	References
Holy basil (*Ocimumtenuiflorum* L.)	Anti-inflammatory effects Antibacterial activities	Prakash and Gupta 2005
Horseradish tree (*Moringaoleifera* Lam.)	Enhancing of wound healing Antibacterial activities Hepatoprotective activity Anti-cancer Antioxidant properties Immunosuppressive activity Anti-inflammatory effects Anti-fungal activity Anti-diabetes mellitus	Jaiswal et al. 2009; Chuang et al. 2007; Mahajan, Mali, and Mehta 2007; Mahajan and Mehta 2010; Atawodi et al. 2010; Budda et al. 2011; Das et al. 2012; Padla et al. 2012
Indian mulberry (*Morindacitrifolia* L.)	Anti-oxidative and anti-inflammatory responses Enhancing of wound healing Induction of apoptosis Anti-diabetes mellitus Beneficial immunomodulation effects UVB-induced transcriptional AP-1 activity	Sang et al. 2003; A. K. Palu et al. 2008; Taşkın et al. 2009; Taşkın et al. 2009; A. Palu et al. 2010; Lin et al. 2013
Common cucumber (*Cucumissativus* L.)	Antioxidant properties Antimicrobial activities	Tang et al. 2010; Kumar et al. 2010
Aloe (*Aloe vera* (L.) Burm. f.)	Antioxidant properties	López et al. 2013

The herb is not recommended for people who have heart disease, asthma, low blood pressure, or leukaemia.

1.2.3 *Clerodendrum petasites* S. Moore (Verbenaceae)

The methanol extract from *Clerodendrum petasites* (Figure 1.4) possesses anti-pyretic and moderate anti-inflammatory activities (Panthong et al. 2003). Many phenolic compounds contained in *Clerodendrum petasites* are known to possess antimicrobial, anti-inflammatory, and anti-oxidant activities; they can be absorbed percutaneously, making them optimal for topical usage (Thitilertdecha, Guy, and Rowan 2014). Roots from white flowers – *Clerodendrum petasites* can help alleviate aphthous ulcers, quench thirst, relieve nausea and vomiting, relieve abscess pain, and relieve coughing from allergies or asthma. Roots from white and red flowers – *Clerodendrum petasites* can also neutralise snake venom (Pichiansunthorn and Jirawong 2013).

FIGURE 1.2 Photograph of *Centella asiatica* showing the round leaves and long stems.

1.2.4 *Curcuma longa* L., *Curcuma domestica* Valeton

Curcuma longa L. (Figure 1.5) contains many types of yellow volatile oils or essential oils. These compounds include aromatic turmerone (60%), which possesses anti-inflammatory and neuroprotective properties (reviewed by B Mythri and M Srinivas Bharath 2012) and induces neural stem cell proliferation *in vitro* and *in vivo* (Hucklenbroich et al. 2014), zingiberene, borneol, and orange-yellowish curcumin (Pichiansunthorn and Jirawong 2013). Due to its colour, curcumin has been

FIGURE 1.3 Photograph of *Rhinacanthus nasutus* showing the oblong leaves 4–10 cm in length, and white flowers.

used for many purposes such as food colouring and textile dyes. *Curcuma longa* L. has a bitter taste but a pleasant smell and can be eaten fresh or dry. It helps purify blood, relieve bloating, gassy stomach and diarrhoea, and treat gastritis. For topical uses, it is ground up and cooked with oil, which is applied to the skin to accelerate wound healing and relieve skin rash or itchiness. Its juice extract also has antiseptic properties to treat fresh wounds, periodontal diseases, gingivitis, and haemorrhagic conjunctivitis (Pichiansunthorn and Jirawong 2013; Termwiset et al. 2013).

FIGURE 1.4 Photograph of *Clerodendrum petasites*.

FIGURE 1.5 Photograph of *Curcuma longa* rhizomes.

REFERENCES

Atawodi, Sunday E, Joy C Atawodi, Gabriel A Idakwo, Beate Pfundstein, Roswitha Haubner, Gerd Wurtele, Helmut Bartsch, and Robert W Owen. 2010. "Evaluation of the Polyphenol Content and Antioxidant Properties of Methanol Extracts of the Leaves, Stem, and Root Barks of Moringa Oleifera Lam." *Journal of Medicinal Food* 13 (3): 710–716.

Babu, T D, G Kuttan, and J Padikkala. 1995. "Cytotoxic and Anti-Tumour Properties of Certain Taxa of Umbelliferae with Special Reference to Centella Asiatica (L.) Urban." *Journal of Ethnopharmacology* 48 (1): 53–57.

Benedec, Daniela, Alina Elena Pârvu, Ilioara Oniga, Anca Toiu, and Brînduoa Tiperciuc. 2007. "Effects of Ocimum Basilicum L. Extract on Experimental Acute Inflammation." *Revista medico-chirurgicala a Societatii de Medici si Naturalisti din Ias* 111 (4): 1065–1069.

B Mythri, R, and M M Srinivas Bharath. 2012. "Curcumin: A Potential Neuroprotective Agent in Parkinson's Disease." *Current Pharmaceutical Design* 18 (1): 91–99.

Bora, Kundan Singh, Shruti Arora, and Richa Shri. 2011. "Role of Ocimum Basilicum L. in Prevention of Ischemia and Reperfusion-Induced Cerebral Damage, and Motor Dysfunctions in Mice Brain." *Journal of Ethnopharmacology* 137 (3): 1360–1365.

Budda, Sirintip, Chaniphun Butryee, Siriporn Tuntipopipat, Anudep Rungsipipat, Supradit Wangnaithum, Jeong-Sang Lee, and Piengchai Kupradinun. 2011. "Suppressive Effects of Moringa Oleifera Lam Pod against Mouse Colon Carcinogenesis Induced by Azoxymethane and Dextran Sodium Sulfate." *Asian Pacific Journal of Cancer Prevention* 12 (12): 3221–3228.

Butt, Masood Sadiq, and M Tauseef Sultan. 2011. "Ginger and Its Health Claims: Molecular Aspects." *Critical Reviews in Food Science and Nutrition* 51 (5): 383–393.

Chuang, Ping-Hsien, Chi-Wei Lee, Jia-Ying Chou, M Murugan, Bor-Jinn Shieh, and Hueih-Min Chen. 2007. "Anti-Fungal Activity of Crude Extracts and Essential Oil of Moringa Oleifera Lam." *Bioresource Technology* 98 (1): 232–236.

Das, Nilanjan, Kunal Sikder, Santinath Ghosh, Bernard Fromenty, and Sanjit Dey. 2012. "Moringa Oleifera Lam. Leaf Extract Prevents Early Liver Injury and Restores Antioxidant Status in Mice Fed with High-Fat Diet." 50 (6): 404–412.

Fathiazad, Fatemeh, Amin Matlobi, Arash Khorrami, Sanaz Hamedeyazdan, Hamid Soraya, Mojtaba Hammami, Nasrin Maleki-Dizaji, and Alireza Garjani. 2012. "Phytochemical Screening and Evaluation of Cardioprotective Activity of Ethanolic Extract of Ocimum Basilicum L.(basil) against Isoproterenol Induced Myocardial Infarction in Rats." *DARU Journal of Pharmaceutical Sciences* 20 (1): 87.

Gnanapragasam, A, K Kumar Ebenezar, V Sathish, P Govindaraju, and T Devaki. 2004. "Protective Effect of Centella Asiatica on Antioxidant Tissue Defense System against Adriamycin Induced Cardiomyopathy in Rats." *Life Sciences* 76 (5): 585–597.

Gohil, Kashmira J, Jagruti A Patel, and Anuradha K Gajjar. 2010. "Pharmacological Review on Centella Asiatica: A Potential Herbal Cure-All." *Indian Journal of Pharmaceutical Sciences* 72 (5): 546.

Hucklenbroich, Joerg, Rebecca Klein, Bernd Neumaier, Rudolf Graf, Gereon Rudolf Fink, Michael Schroeter, and Maria Adele Rueger. 2014. "Aromatic-Turmerone Induces Neural Stem Cell Proliferation in Vitro and in Vivo." *Stem Cell Research & Therapy* 5 (4): 100.

Jaiswal, Dolly, Prashant Kumar Rai, Amit Kumar, Shikha Mehta, and Geeta Watal. 2009. "Effect of Moringa Oleifera Lam. Leaves Aqueous Extract Therapy on Hyperglycemic Rats." *Journal of Ethnopharmacology* 123 (3): 392–396.

Kaurinovic, Biljana, Mira Popovic, Sanja Vlaisavljevic, and Svetlana Trivic. 2011. "Antioxidant Capacity of Ocimum Basilicum L. and Origanum Vulgare L. Extracts." *Molecules* 16 (9): 7401–7414.

Kaya, Ilhan, Nazife Yigit, and Mehlika Benli. 2008. "Antimicrobial Activity of Various Extracts of Ocimum Basilicum L. and Observation of the Inhibition Effect on Bacterial Cells by Use of Scanning Electron Microscopy." *African Journal of Traditional, Complementary and Alternative Medicines* 5 (4): 363–369.

Kumar, D, S Kumar, J Singh, B D Vashistha, and N Singh. 2010. "Free Radical Scavenging and Analgesic Activities of Cucumis Sativus L. Fruit Extract." *Journal of Young Pharmacists* 2 (4): 365–368.

Lin, Yi-Ling, Yuan-Yen Chang, Deng-Jye Yang, Bor-Show Tzang, and Yi-Chen Chen. 2013. "Beneficial Effects of Noni (Morinda Citrifolia L.) Juice on Livers of High-Fat Dietary Hamsters." *Food Chemistry* 140 (1): 31–38.

López, Aroa, Miguel Suárez de Tangil, Orestes Vega-Orellana, Ana S Ramírez, and Milagros Rico. 2013. "Phenolic Constituents, Antioxidant and Preliminary Antimycoplasmic Activities of Leaf Skin and Flowers of Aloe Vera (L.) Burm. F.(syn. A. Barbadensis Mill.) from the Canary Islands (Spain)." *Molecules* 18 (5): 4942–4954.

Mahajan, Shailaja G, Ravindra G Mali, and Anita A Mehta. 2007. "Protective Effect of Ethanolic Extract of Seeds of Moringa Oleifera Lam. against Inflammation Associated with Development of Arthritis in Rats." *Journal of Immunotoxicology* 4 (1): 39–47.

Mahajan, Shailaja G, and Anita A Mehta. 2010. "Immunosuppressive Activity of Ethanolic Extract of Seeds of Moringa Oleifera Lam. in Experimental Immune Inflammation." *Journal of Ethnopharmacology* 130 (1): 183–186.

Marzouk, Amani M. 2009. "Hepatoprotective Triterpenes from Hairy Root Cultures of Ocimum Basilicum L." *Zeitschrift Für Naturforschung C* 64 (3–4): 201–209.

Munasinghe, M A A K, C Abeysena, I S Yaddehige, T Vidanapathirana, and K P B Piyumal. 2011. "Blood Sugar Lowering Effect of Coccinia Grandis (L.) J. Voigt: Path for a New Drug for Diabetes Mellitus." Experimental Diabetes Research 2011. Hindawi Publishing Corporation.

Murakami, Akira, Guanxin Gao, Oe Kyung Kim, Mitsuo Omura, Masamichi Yano, Chihiro Ito, Hiroshi Furukawa, Suratwadee Jiwajinda, Koichi Koshimizu, and Hajime Ohigashi. 1999. "Identification of Coumarins from the Fruit of Citrus Hystrix DC as Inhibitors of Nitric Oxide Generation in Mouse Macrophage RAW 264.7 Cells." *Journal of Agricultural and Food Chemistry* 47 (1): 333–339.

Ogra, Yasumitsu, Kazuya Ishiwata, Yuji Iwashita, and Kazuo T Suzuki. 2005. "Simultaneous Speciation of Selenium and Sulfur Species in Selenized Odorless Garlic (Allium Sativum L. Shiro) and Shallot (Allium Ascalonicum) by HPLC–inductively Coupled Plasma-(Octopole Reaction System)-Mass Spectrometry and Electrospray Ionization-Tandem Mass Spectrometry." *Journal of Chromatography A* 1093 (1): 118–125.

Padla, Eleanor P, Ludivina T Solis, Ruel M Levida, Chien-Chang Shen, and Consolacion Y Ragasa. 2012. "Antimicrobial Isothiocyanates from the Seeds of Moringa Oleifera Lam." *Zeitschrift Für Naturforschung C* 67 (11–12): 557–564.

Palu, Afa K, Anne Hirazumi Kim, Brett J West, Shixin Deng, Jarakae Jensen, and Leland White. 2008. "The Effects of Morinda Citrifolia L.(noni) on the Immune System: Its Molecular Mechanisms of Action." *Journal of Ethnopharmacology* 115 (3): 502–506.

Palu, Afa, Chen Su, Bing-Nan Zhou, Brett West, and Jarakae Jensen. 2010. "Wound Healing Effects of Noni (Morinda Citrifolia L.) Leaves: A Mechanism Involving Its PDGF/A2A Receptor Ligand Binding and Promotion of Wound Closure." *Phytotherapy Research* 24 (10): 1437–1441.

Panthong, A, D Kanjanapothi, T Taesotikul, T Wongcome, and V Reutrakul. 2003. "Anti-Inflammatory and Antipyretic Properties of Clerodendrum Petasites S. Moore." *Journal of Ethnopharmacology* 85 (1): 151–156.

Parameshwaraiah, S, and H G Shivakumar. 1998. "Evaluation of Topical Formulations of Aqueous Extract of Centella Asiatica on Open Wounds in Rats." *Indian Journal of Experimental Biology* 36 (6): 569–572.

Pari, Leelavinothan, Daniel Tewas, and Juergen Eckel. 2008. "Role of Curcumin in Health and Disease." *Archives of Physiology and Biochemistry* 114 (2): 127–149.

Pichiansunthorn, Chayan, and Vichian Jirawong. 2013. "[The Object of Botanical Medicine]." In [Thailand Traditional Pharmacy Guide], 2nd ed. Bangkok: Amarin Publisher.

Piyachaturawat, Pawinee, Thirayudh Glinsukon, and Aporn Chanjarunee. 1985. "Antifertility Effect of Citrus Hystrix DC." *Journal of Ethnopharmacology* 13 (1): 105–110.

Prakash, P, and Neelu Gupta. 2005. "Therapeutic Uses of Ocimum Sanctum Linn (Tulsi) with a Note on Eugenol and Its Pharmacological Actions: A Short Review." *Indian Journal of Physiology and Pharmacology* 49 (2): 125.

Sang, Shengmin, Guangming Liu, Kan He, Nanqun Zhu, Zigang Dong, Qunyi Zheng, Robert T Rosen, and Chi-Tang Ho. 2003. "New Unusual Iridoids from the Leaves of Noni (Morinda Citrifolia L.) Show Inhibitory Effect on Ultraviolet B-Induced Transcriptional Activator Protein-1 (AP-1) Activity." *Bioorganic & Medicinal Chemistry* 11 (12): 2499–2502.

Saralamp, Promjit, Wongsatit Chuakul, Rungravi Temsiririrkkul, and Terry Clayton. 1996. *Medicinal Plants in Thailand.* Vol. 1. Department of Pharmaceutical Botany, Faculty of Pharmacy, Mahidol University, Bangkok, 196–204.

Satheesh, L Shilpa, and K Murugan. 2011. "Antimicrobial Activity of Protease Inhibitor from Leaves of Coccinia Grandis (L.) Voigt." NISCAIR-CSIR, India.

Shamnan, Dharmanit. 2013. *Thailand Medicinal – Herb Bible.* Bangkok: Thaiquality book.

Singh, Geetu, Prasoon Gupta, Preeti Rawat, Anju Puri, Gitika Bhatia, and Rakesh Maurya. 2007. "Antidyslipidemic Activity of Polyprenol from Coccinia Grandis in High-Fat Diet-Fed Hamster Model." *Phytomedicine* 14 (12): 792–798.

Tang, Jing, Xiangjie Meng, Hao Liu, Jianglin Zhao, Ligang Zhou, Minghua Qiu, Xianming Zhang, Zhu Yu, and Fuyu Yang. 2010. "Antimicrobial Activity of Sphingolipids Isolated from the Stems of Cucumber (Cucumis Sativus L.)." *Molecules* 15 (12): 9288–9297.

Taşkın, Elif İlkay, Kadriye Akgün-Dar, Ayşegül Kapucu, Esma Osanç, Hüsniye Doğruman, Hakan Eraltan, and Engin Ulukaya. 2009. "Apoptosis-inducing Effects of Morinda Citrifolia L. and Doxorubicin on the Ehrlich Ascites Tumor in Balb-c Mice." *Cell Biochemistry and Function* 27 (8): 542–546.

Termwiset, Porntip, Bussaraporn Thanaselungkul, Thanatip Chimpae, and Kwanruen Chantee. 2013. *Collection of Properties of Thai Herbs.* Bangkok: Affairs Printing Office of the War Veterans Organization.

Thitilertdecha, Premrutai, Richard H Guy, and Michael G Rowan. 2014. "Characterisation of Polyphenolic Compounds in Clerodendrum Petasites S. Moore and Their Potential for Topical Delivery through the Skin." *Journal of Ethnopharmacology* 154 (2): 400–407.

Umamaheswari, M, and T K Chatterjee. 2008. "In Vitro Antioxidant Activities of the Fractions of Coccinia Grandis L. Leaf Extract." *African Journal of Traditional, Complementary and Alternative Medicines* 5 (1): 61–73.

Veerendra Kumar, M H, and Y K Gupta. 2003. "Effect of Centella Asiatica on Cognition and Oxidative Stress in an Intracerebroventricular Streptozotocin Model of Alzheimer's Disease in Rats." *Clinical and Experimental Pharmacology and Physiology* 30 (5–6): 336–342.

Wongmekiat, O, N Leelarugrayub, and K Thamprasert. 2008. "Beneficial Effect of Shallot (Allium Ascalonicum L.) Extract on Cyclosporine Nephrotoxicity in Rats." *Food and Chemical Toxicology* 46 (5): 1844–1850.

2 Methods for Extracting Medicinal Compounds from Thai Herbs

James M. Brimson, Sirikalaya Brimson, and Tewin Tencomnao

2.1 INTRODUCTION

Many currently prescribed drugs have been either directly derived from or inspired by natural products (Atanasov et al., 2015). Clay tablets in cuneiform from Mesopotamia (2600 B.C.) contain the earliest documented records of natural products. These tablets describe oils derived from *Cupressus sempervirens* (Cypress) and *Commiphora* species (myrrh), both of which are still employed today for treating coughs, colds, and inflammation (Cragg and Newman, 2005). All natural products begin as mixtures with other compounds from the natural source, often very complex mixtures, from which the product of interest must be isolated and purified.

Herbal extraction for medicinal purposes can be achieved using a variety of methods, depending on the type of medicine (cream, liquid, gel, etc.). In traditional medicine prior to knowledge of extraction, herbs tended to be used fresh or dried and may have been boiled for drinking (Sentkowska et al., 2016) (like a tea) or a compression made with mashed herbs placed on the affected area (Nikam et al., 2023). Following a better understanding of extraction, crude extracts were applied for therapeutic purposes. In some cases, the crude extract is more effective than a pure substance due to synergistic effects of multiple substances in particular herbs (Rasoanaivo et al., 2011). Whereas in other cases, however, the crude extract may have antagonistic effects from mixed substances or low efficacy due to the dilution of an active compound. Therefore, purification and isolation with known chemical structures are also required and can be used as markers in many herbs.

When extracting natural products from herbs and plants, the selection of the solvent and the extraction method is key to getting a high yield and collecting the compounds you are interested in (Azmin et al., 2015). When extracting with multiple solvents, the solvent with the lowest polarity is used first, followed sequentially by solvents with the highest polarity, allowing for the highest yield of extracted compounds and providing a certain amount of separation of the different compounds

DOI: 10.1201/9781003508724-2

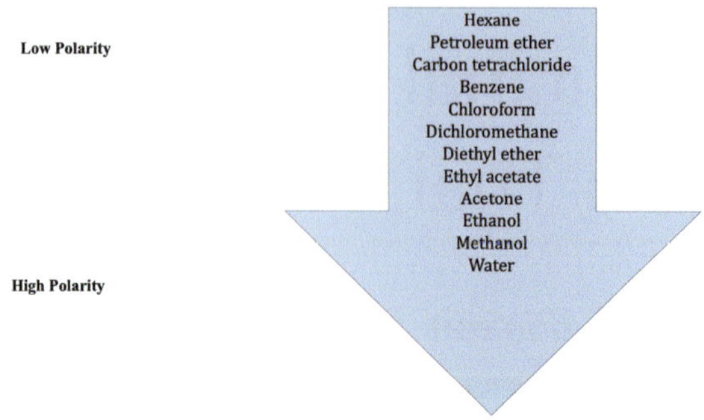

FIGURE 2.1 Solvents from low to high polarity.

depending on their solubility in each solvent. The polarity of various solvents is shown in Figure 2.1.

It is important to select the correct part of the herb (root, leaf, stem, etc.), as each part may contain differing amounts and types of compounds. Equally important is understanding that the time of year or the plant's growing conditions and environment can also impact the yield of extract. For example, clove or *Eugenia aromaticam* produces the maximum oil level when their flowers are yet to bloom.

2.2 DRYING OF HERBS

Herbs are dried to stop enzymes from working and to prevent bacterial activity so that the active compounds can be preserved for long periods of time.

Air drying – this method is suitable for herbs containing heat/sunlight-labile substances.

Sun drying – this method is not good for herbs containing oil and requires a significant amount of sunlight where there is less risk of rain. If there are not good conditions for sun drying, oven drying will be more suitable.

Hot air oven – the temperature in the oven should not be higher than 50°C.

Freeze drying – this method can preserve easily degradable substances. Extraction processes can be divided into cold and hot processes. The cold processes are suitable for extracting compounds that are heat labile (Nurhaslina et al., 2022).

2.3 COLD PROCESSES

2.3.1 Maceration

Maceration (Figure 2.2) is one of the simplest and most used extraction techniques. In much the same way as making a cup of tea, where tea is extracted from tea leaves

FIGURE 2.2 Extraction of *Rhinacanthus nasutus* root (from left to right) in water, 80% ethanol, 100% ethanol and hexane.

by allowing them to soak in water for a prescribed amount of time, herbs are dried and placed in a sealed vessel with the solvent of choice, commonly water (care should be taken to avoid contamination and bacterial growth), ethanol, methanol, or hexane. The herb is then incubated at room temperature for a period of two to seven days, with agitation, before the solvent is then filtered and collected. To ensure complete extraction, the solid material can then be re-extracted with more solvent. Once the extraction is complete, the solvent containing the extracted materials is concentrated, usually resulting in a pigmented waxy solid (Ćujić et al., 2016).

2.3.2 Percolation

Percolation (Figure 2.3) is a very simple extraction technique that involves the slow movement of solvent over the herb before filtration and collection of the extract. The herb is suspended on a perforated base, above a filter in a vessel where solvent can be poured on top. The solvent soaks through the herb and slowly drips into the collection vessel, taking the extracted compounds with it. If desired, it is possible to heat the vessel containing the herb to facilitate extraction. This method of extraction may be familiar to those who use a coffee percolator for coffee consumption, where hot water is passed over a filter paper containing the ground coffee and collected in a jug below (Hidayat et al., 2021).

2.3.3 Soxhlet Extraction

The Soxhlet extraction process (Figure 2.4) uses a heated solvent but is a cold extraction process, as the solvent is cooled before contact with the dried herb; thus, the extraction is cold. However, this process is still not applicable to extracts that may

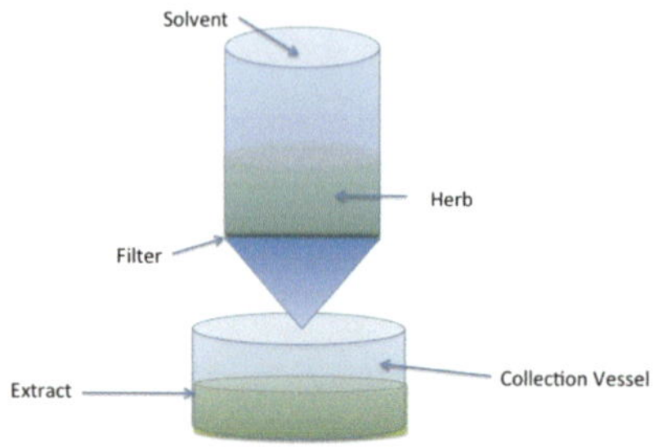

FIGURE 2.3 Diagram of apparatus used in percolation.

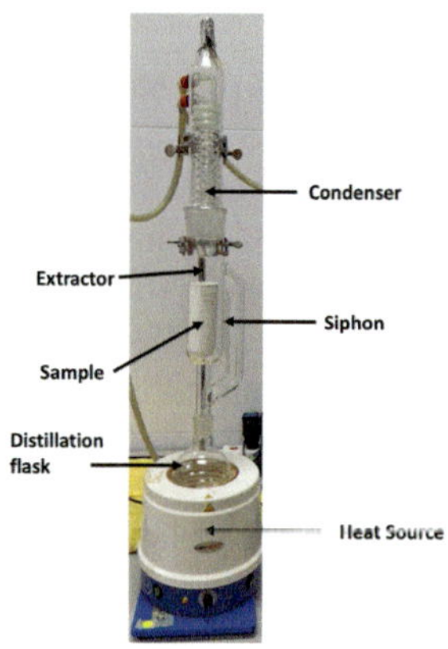

FIGURE 2.4 Photograph of Soxhlet apparatus.

be heat labile (López-Bascón et al., 2020). A pool of solvent in the distillation flask is gently heated, causing it to evaporate and travel up to the condenser, where the solvent is cooled and condensed, causing it to drip down onto the herb sample. As the solvent fills up over the herb compounds are extracted until the solvent level reaches the syphon. At this point, the solvent containing the extract is syphoned back down into the pool of solvent in the distillation flask. This process is repeated until the extract is concentrated enough that it no longer stays in solution. The Soxhlet has the advantage over maceration and percolation, as a smaller amount of solvent can be recycled and used to extract more compounds from the herb (De Castro et al., 1998).

2.3.4 Supercritical CO_2 Extraction

Supercritical CO_2 extraction (Figure 2.5) is a popular method used by coffee manufacturers to extract caffeine from coffee, in the production of decaffeinated coffee, and to isolate the caffeine for resale to pharmaceutical or beverage companies (Tello et al., 2011). CO_2 under normal conditions is a gas or can be a solid if significantly cooled (dry ice); however, under pressure and heat (above 31°C and 1071 psi), CO_2 can behave as a supercritical liquid (in between a liquid and gas).

Supercritical CO_2 is passed over the herb, where compounds dissolve in the supercritical CO_2 inside a pressure vessel. The liquid CO_2 is then passed into a collection vessel with a lower pressure than the extraction vessel, in which the extract will precipitate. The solubility of the compounds in the supercritical CO_2 varies depending on the density of the liquid; therefore, the extracted compounds can be fractionated and specific compounds selected based on their solubility at different pressures of the liquid CO_2 (Machmudah et al., 2011).

One of the main advantages of this extraction method is its selectivity, as the properties of the supercritical fluid can be altered depending on the pressure and temperature, with caffeine extraction from coffee beans being a good example. By selecting the correct temperature and pressure, only the caffeine is extracted, leaving behind all compounds that give coffee its distinctive taste and smell (Zakharenko et al., 2020). Volatile oils can be extracted from plants at low pressures (1450 psi), and lipids can be extracted at higher pressures. Another advantage of this process is its speed; unlike methods such as maceration and Soxhlet extraction (which can take days to fully extract compounds), supercritical fluids can extract compounds in as little as 10 to 60 minutes. The increase in speed is brought about by the fact that supercritical fluids lack surface tension and have low viscosities; furthermore, diffusivities are much faster in supercritical fluids (Pourmortazavi et al., 2007). A major disadvantage of supercritical CO_2 extraction is the requirement for such high

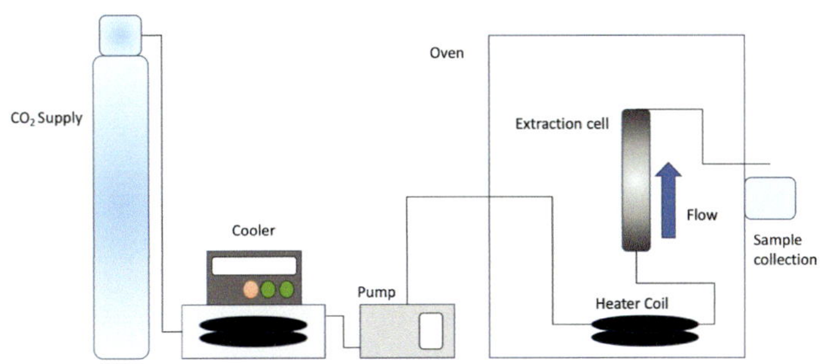

FIGURE 2.5 Diagram of super critical CO_2 equipment.

pressures, which can increase cost when compared to other liquid extraction methods. CO_2, being non-polar, has limited dissolving power per se so it cannot always be used as the sole solvent, and other solvents such as ethanol are required to be used as modifiers, which somewhat negates a number of the benefits (such as the ease of recovering extracted compounds) of using a solvent that is a gas at room temperature and pressure.

2.4 HOT PROCESSES

Hot processes are applicable when the compounds that are to be isolated are not heat labile. Using heat in the process speeds up the rate at which molecules can diffuse, increasing the yield and reducing the time required for the extraction.

2.4.1 Decoction or Boiling

Herbs are boiled with solvent directly (Figure 2.6), and the extracted liquid can be consumed or evaporated to keep the extract as a powder. Freeze drying is often used to remove water from the desired extract (Zhang et al., 2017).

2.4.2 Infusion

Herbs are immersed in warm, not boiling, solvent for a specified amount of time. The drink "tea" is a common example of an infusion that is used in homes worldwide (Figure 2.7) to extract the compounds from dried tea leaves that give tea its distinctive smell and taste (Martono et al., 2020).

FIGURE 2.6 Photograph of a traditional Thai clay pot for boiling herbs.

FIGURE 2.7 Photograph of a teapot, commonly used for steeping herbs (usually tea leaves, although other herbal mixes are often used) in people's own homes.

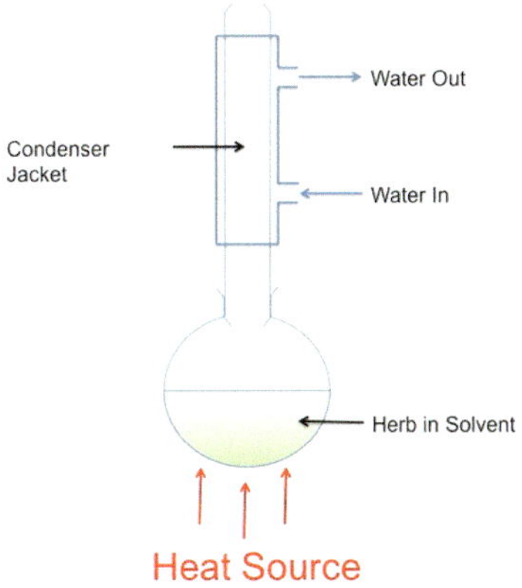

FIGURE 2.8 Diagram of reflux apparatus.

2.4.3 Reflux Extraction

Herbs are boiled in a fixed volume of solvent (Figure 2.8). The evaporated solvent is condensed in a condenser and dropped back into the boiling vessel. Active compounds or compounds of interest are more readily dissolved with the heat, and the condenser prevents the solvent volume from reducing, allowing a greater yield than boiling alone. However, the volume of solvent limits the yield, which is different from Soxhlet extraction where, through the separation of the extraction vessel and collection vessel, the solvent does not become saturated (Ariff et al., 2020).

2.4.4 Distillation

Distillation is applicable for the extraction of volatile oils and aromatic compounds from plants. There are various distillation methods that are suitable for extracting plant volatile oils. The use of these methods depends on the amount of oil, how heat labile the substance is, where the oil is found in the plant, and the final use for the compound (e.g., volatile, ointment, or cream).

2.4.5 Steam Distillation

This method is good for plants that collect oil in the glandular trichome or oil reservoir. Plants with oil reservoirs include basil, peppermint, orange peels, and lime leaves. Plants can be spread on a sieve, grate, or screen over the water in a closed system with a condenser or in a separate vessel connected to the steam generating vessel, as in Figure 2.9. When water is boiled, steam is passed over and through the plants (Figure 2.9), causing the release of oil from the glandular trichomes or oil reservoir (Chemat and Boutekedjiret, 2015). The volatile oil evaporates and mixes with the steam, which carries it to the condenser, where both water and volatile oil condense and can be collected as separated layers of water and oil. This method is convenient, fast, and inexpensive.

2.4.6 Hydro-distillation

This method is similar to steam distillation, with the exception being that plants are boiled with the water directly. Therefore, the herb being extracted gets hotter than that in a method such as steam distillation, meaning this method is more suitable for active compounds that do not decompose when heated. The condensed oil and water are collected from the condenser and form separate layers, (Figure 2.10) making the collection of the oil very straight forward (Denny, 1989).

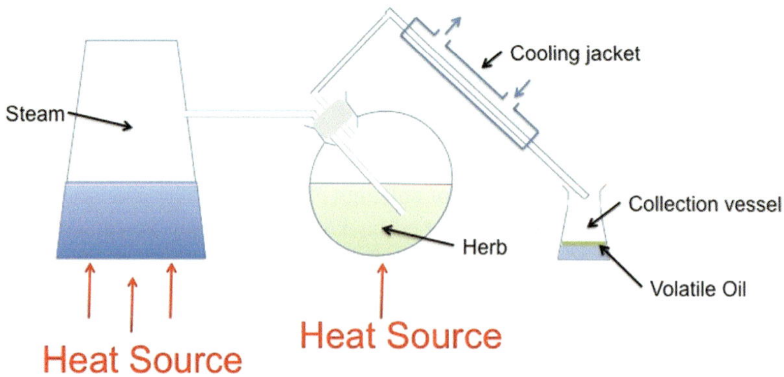

FIGURE 2.9 Diagram of steam distillation.

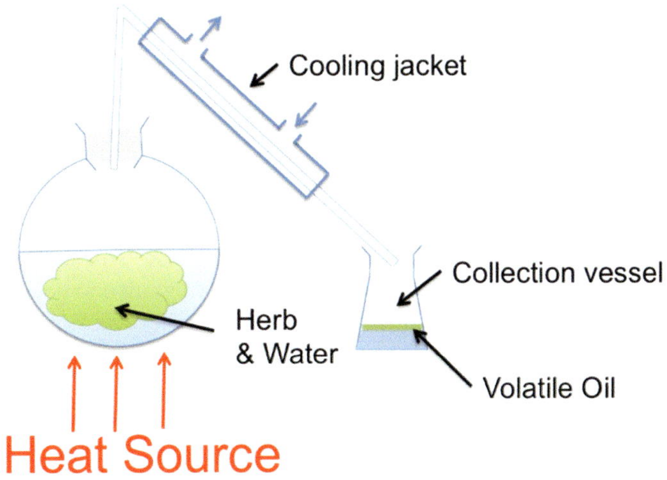

FIGURE 2.10 Diagram of hydro-distillation.

2.4.7 Destructive Distillation

This method can only be used with a select few types of plants, one example being the extraction of pine oil (a byproduct of charcoal production). The plant is chopped into small pieces and heated in a metal container at a high temperature in a low oxygen environment. The oil is collected as it drips out of the bottom of the container. Some of the oil may be lost as it burns from the heat flame. The oil product is usually black or dark in colour and suitable for industrial uses, e.g., industrial disinfectant (Rao et al., 2019).

There are other ways to extract the volatile oil apart from distillation, including extraction with solvent and extraction with fat.

2.4.8 Oil Extraction with Solvent

The required properties of the solvent for this method include high purity and being highly volatile and hydrophobic. Herbs are soaked in the solvent for a period of time before the solvent is separated from the plants by filtration. The filtrated solvent is then evaporated using distillation at low temperature under pressure. The concentrated solvent is called "concrete," which will be washed with alcohol many times to get rid of contaminants. The final oil product is called an "absolute" (Roohinejad et al., 2017).

An example of this method involves using "supercritical carbon dioxide extraction" (described earlier). This method is costly but has some advantages in the properties of the product, including being easily separable and leaving no scent, taste, toxicity, or flammability in the product. Other solvents may cause dark colours in the products and leave behind contaminants.

2.4.9 Enfleurage

This method is suitable for extracting oil from the parenchyma in petals, such as those from roses and jasmines, in which their fragrances are well preserved. Normally, pig, cow, or sheep fat is spread on trays where petals are evenly lined up on top in a closed system for 24 hours. Petals are removed and replaced with new petals every 24 hours until petal oil is saturated in fat (pomade). Petal oil is then extracted from pomade using alcohol (extrait or absolute de pomade). The remaining fat can be used to produce soaps as it retains a certain amount of fragrance. This method was once popular in the fragrance/perfume industry as fragile fragrant compounds could be extracted at lower temperatures; however, these methods have mostly superseded by supercritical CO_2 extraction (Hanif et al., 2019).

2.5 PURIFICATION METHODS

2.5.1 Phase Separation

Phase separation refers to the formation of two distinct phases from an initially uniform mixture. The most prevalent form of phase separation occurs between two liquids that are immiscible, such as oil and water. By mixing the solvent containing the crude herbal extract with another solvent, some of the compounds that are more soluble in the second solvent will be taken with that solvent when the two immiscible liquids are allowed to separate, thus removing some compounds but not others.

2.5.2 Liquid Chromatography

Liquid column chromatography is a widely employed technique for separating compounds within a crude herbal extract based on their distinct chemical properties. In this method, the extract is initially dissolved in a liquid solvent known as the mobile phase, which is then passed through a column filled with a stationary phase. The stationary phase, often composed of a solid support material, interacts selectively with the components of the extract. As the mobile phase flows through the column, the individual compounds within the herbal extract experience differential interactions with the stationary phase, causing them to separate (Tistaert et al., 2011). The extent of separation is influenced by factors such as molecular size, polarity, and affinity for the stationary phase. By adjusting these parameters, liquid column chromatography allows for the isolation and purification of specific compounds from the complex mixture, providing a valuable tool in the analysis and extraction of bioactive substances from herbal sources (Dejaegher et al., 2010).

2.5.3 HPLC

High-performance liquid chromatography (HPLC) stands as a powerful method for the separation of compounds within a crude herbal extract, offering enhanced efficiency and precision (Liu et al., 2007). In HPLC, the sample is dissolved in a liquid solvent, serving as the mobile phase, and is forced through a column filled with a stationary phase at high pressure. The stationary phase, often composed of fine particles, allows for more rapid and efficient separation compared to traditional liquid chromatography methods. As the sample travels through the column, individual compounds interact with the stationary phase based on factors such as polarity, size, and chemical structure. The high pressure employed in HPLC facilitates faster and more accurate separations, enabling the isolation of specific compounds from the complex herbal mixture. HPLC is widely utilised in the pharmaceutical and analytical industries due to its ability to provide precise and reproducible results, making it an invaluable tool for identifying and quantifying bioactive constituents in herbal extracts (Castro-Rios et al., 2020).

REFERENCES

Ariff, M. A. M., & Abdullah, N. (2020). Optimization of reflux extraction for cat's whiskers leaves extract using response surface methodology. *Chemical Industry and Chemical Engineering Quarterly*, 26(1), 49–57.

Atanasov, A. G., Waltenberger, B., Pferschy-Wenzig, E. M., et al. 2015. Discovery and resupply of pharmacologically active plant-derived natural products: A review. *Biotechnology Advances*, 33(8), 1582–1614.

Azmin, S. N. H. M., Yunus, N. A., Mustaffa, A. A., Wan Alwi, S. R., & Chua, L. S. (2015). A framework for solvent selection based on herbal extraction process design. *Journal of Engineering Science and Technology*, 1, 25–34.

Castro-Ríos, R., Melchor-Martínez, E. M., Solís-Cruz, G. Y., Rivas-Galindo, V. M., Silva-Mares, D. A., & Cavazos-Rocha, N. C. (2020). HPLC method validation for Jatropha dioica extracts analysis. *Journal of Chromatographic Science*, 58(5), 445–453.

Chemat, F., & Boutekedjiret, C. (2015). Extraction//steam distillation. Reference Module in Chemistry, Molecular Sciences and Chemical Engineering; Elsevier: Amsterdam, The Netherlands, 1–12.

Cragg, G. M., & Newman, D. J. (2005). Biodiversity: A continuing source of novel drug leads. *Pure and Applied Chemistry*, 77(1), 7–24.

Ćujić, N., Šavikin, K., Janković, T., Pljevljakušić, D., Zdunić, G., & Ibrić, S. (2016). Optimization of polyphenols extraction from dried chokeberry using maceration as traditional technique. *Food Chemistry*, 194, 135–142.

De Castro, M. L., & Garcıa-Ayuso, L. E. (1998). Soxhlet extraction of solid materials: An outdated technique with a promising innovative future. *Analytica chimica acta*, 369(1–2), 1–10.

Dejaegher, B., Alaerts, G., & Matthijs, N. (2010). Methodology to develop liquid chromatographic fingerprints for the quality control of herbal medicines. *Acta Chromatographica*, 22(2), 237–258.

Denny, E. F. K. (1989). Hydro-distillation of oils from aromatic herbs. *Perfum Flavor*, 14(5763), 42.

Hanif, M. A., Nisar, S., Khan, G. S., Mushtaq, Z., & Zubair, M. (2019). Essential oils. Essential Oil Research: Trends in Biosynthesis, Analytics, Industrial Applications and Biotechnological Production, 3–17.

Hidayat, R., & Wulandari, P. (2021). Methods of extraction: Maceration, percolation and decoction. *Eureka Herba Indonesia*, 2(1), 68–74.

Liu, M., Li, Y., Chou, G., Cheng, X., Zhang, M., & Wang, Z. (2007). Extraction and ultra-performance liquid chromatography of hydrophilic and lipophilic bioactive components in a Chinese herb Radix Salviae Miltiorrhizae. *Journal of Chromatography A*, 1157(1–2), 51–55.

López-Bascón, M. A., & De Castro, M. L. (2020). Soxhlet extraction. In *Liquid-Phase Extraction* (pp. 327–354). Elsevier.

Machmudah, S., Kitada, K., Sasaki, M., Goto, M., Munemasa, J., & Yamagata, M. (2011). Simultaneous extraction and separation process for coffee beans with supercritical CO2 and water. *Industrial & Engineering Chemistry Research*, 50(4), 2227–2235.

Martono, Y., & Muninggar, J. (2020, October). Optimization infusion condition on flavonoid and antioxidant activity of herb beverages using response surface methodology. In IOP Conference Series: Materials Science and Engineering (Vol. 959, No. 1, p. 012029). IOP Publishing.

Nikam, A., Thomas, A., Giram, P., Nagore, D., & Chitlange, S. (2023). Herbal-based dressings in wound management. *Current Diabetes Reviews*, 19(4), 46–58.

Nurhaslina, C. R., Bacho, S. A., & Mustapa, A. N. (2022). Review on drying methods for herbal plants. *Materials Today: Proceedings*, 63, S122–S139.

Pourmortazavi, S. M., & Hajimirsadeghi, S. S. (2007). Supercritical fluid extraction in plant essential and volatile oil analysis. *Journal of Chromatography A*, 1163(1–2), 2–24.

Rao, H., Li, P., Wu, H., Liu, C., Peng, W., & Su, W. (2019). Simultaneous determination of six compounds in destructive distillation extracts of hawthorn seed by GC-MS and evaluation of their antimicrobial activity. *Molecules*, 24(23), 4328.

Rasoanaivo, P., Wright, C. W., Willcox, M. L., & Gilbert, B. (2011). Whole plant extracts versus single compounds for the treatment of malaria: synergy and positive interactions. *Malaria Journal*, 10(1), 1–12.

Roohinejad, S., Koubaa, M., Barba, F. J., Leong, S. Y., Khelfa, A., Greiner, R., & Chemat, F. (2017). Extraction methods of essential oils from herbs and spices. *Essential Oils in Food Processing: Chemistry, Safety and Applications*, 1, 21–55.

Sentkowska, A., Biesaga, M., & Pyrzynska, K. (2016). Effects of brewing process on phenolic compounds and antioxidant activity of herbs. *Food Science and Biotechnology*, 25, 965–970.

Tello, J., Viguera, M., & Calvo, L. (2011). Extraction of caffeine from Robusta coffee (Coffea canephora var. Robusta) husks using supercritical carbon dioxide. *The Journal of Supercritical Fluids*, 59, 53–60.

Tistaert, C., Dejaegher, B., & Vander Heyden, Y. (2011). Chromatographic separation techniques and data handling methods for herbal fingerprints: A review. *Analytica chimica acta*, 690(2), 148–161.

Zakharenko, A., Romanchenko, D., Thinh, P. D., Pikula, K., Hang, C. T. T., Yuan, W., ... Golokhvast, K. (2020). Features and advantages of supercritical CO2 extraction of sea cucumber Cucumaria frondosa japonica Semper, 1868. *Molecules*, 25(18), 4088.

Zhang, X. F., Yang, J. L., Chen, J., & Shi, Y. P. (2017). Optimization of a decoction process for an herbal formula using a response surface methodology. *Journal of AOAC International*, 100(6), 1776–1784.

3 Thai Medicinal Uses of Agarwood (*Aquilaria crassna*)

Apinan Senabunyarith, Phaniendra Alugoju, Madugula Shabareesh, Siriporn Chuchawankul, James M. Brimson, and Tewin Tencomnao

3.1 INTRODUCTION

Aquilaria spp., also known as Agarwood (Europe), Aloeswood (Singapore), Eaglewood (USA), Gaharu (Indonesia), Oudh (Arabic), Tram (Vietnam), Jinko (Japan), and Chen Xiang (Chinese) (Y. Z. H.-Y. Hashim, P. G. Kerr, P. Abbas, & H. Mohd Salleh, 2016), is widely regarded as an exceptional natural incense and has been integral to cultural, religious, and medicinal purposes in various countries. The Aquilaria tree genus comprises 21 recognised species (refer to Table 3.1). In Thailand, five species are identified, one of which is imported from China, called "Krissana China" or *Aquilaria subintegra*. The Aquilaria species found in Thailand are briefly described below.

a) *Aquilaria crassna* Pierre ex Lecomte, known locally in Thailand as "Krissana," is found in Thailand's eastern, northeastern, northern, and southeastern regions. It's a medium-sized deciduous tree growing to about 10–30 metres tall.
b) *Aquilaria malaccensis*, referred to as "Maihom," is distributed in the southern and eastern parts of Thailand, particularly in the lower regions. It can grow up to around 40 metres in height.
c) *Aquilaria hirta*, known as "Ja Nae," is found in southern Thailand, primarily in Prachuap Khiri Khan. This deciduous tree is small, reaching about 15 metres in height.
d) *Aquilaria subintegra*, locally known as "Gahar," is found in the southern region of Thailand. It is a relatively small tree, growing to a height of 2–5 metres.
e) *Aquilaria rugosa* is a medium-sized deciduous tree, typically 35 metres tall. It is found in the northern areas of Thailand.

DOI: 10.1201/9781003508724-3

TABLE 3.1
Lists the Names of Aquilaria Tree Genus Species Worldwide ("The Plant List", 2013)

Order	Botanical Name	Geographic Distribution Area
1	*Aquilaria apiculata* Merr.	Philippines
2	*Aquilaria baillonii* Pierre ex Lecomte	Cambodia
3	*Aquilaria banaense* P.H.Hô	Vietnam
4	*Aquilaria beccariana* Tiegh.	Borneo, Sarawak, Brunei, Sumatra
5	*Aquilaria brachyantha* (Merr.) Hallier f.	Philippines
6	*Aquilaria citrinicarpa* (Elmer) Hallier f.	Philippines
7	*Aquilaria crassna* Pierre ex Lecomte	Laos, Cambodia, Vietnam, Thailand
8	*Aquilaria cumingiana* (Decne.) Ridl.	Philippines
9	*Aquilaria decemcostata* Hallier f.	Philippines
10	*Aquilaria filaria* (Oken) Merr.	Philippines
11	*Aquilaria hirta* Ridl.	Coastal areas of Thailand and Malaysia (Southern Thailand)
12	*Aquilaria khasiana* Hallier f.	India
13	*Aquilaria malaccensis* Lam.	India, Myanmar, Thailand, Malaysia, Sumatra, Borneo, Philippines
14	*Aquilaria microcarpa* Baill.	Borneo, Sarawak, Sumatra, Brunei
15	*Aquilaria parvifolia* (Quisumb.) Ding Hou	Philippines
16	*Aquilaria rostrata* Ridl.	Malaysia
17	*Aquilaria rugosa* K.Le-Cong & Kessler	Vietnam and Thailand
18	*Aquilaria sinensis* (Lour.) Spreng.	China (Hong Kong, Hainan)
19	*Aquilaria subintegra* Ding Hou	Southern Thailand
20	*Aquilaria urdanetensis* (Elmer) Hallier f.	Philippines
21	*Aquilaria yunnanensis* S.C.Huang	China (Yunnan)

3.1.1 Botanical Description

Scientific name: *Aquilaria crassna* Pierre ex Lecomte
Local name: Krassna
Characteristics: Forest Research and Development Office, 2557

Description: It is a deciduous tree of medium size (Figure 3.1), reaching heights of 10–30 metres. The crown is either low or cup-shaped. The trunk is straight, and

FIGURE 3.1 The characteristic of the stem/trunk (A), flower (B), fruit (C), seed (D), and leaf (E) of *Aquilaria crassna* Pierre ex Lecomte. Images kindly provided by Mr Supakorn Yuenyongwannachot.

often, the base has small, low, or cup-shaped buttresses. As the tree ages, the outer bark of the trunk becomes smooth, with a soft grey colour marked by pores for air circulation. The inner bark of the trunk is white, with a sticky yellowish tint covering the young branches. The young branches are covered with soft hairs.

Bark: In mature trees, the outer bark is smooth, grey, or marked with fissures, while the inner bark is yellowish-white and slightly sticky, with soft hair covering it.

Leaves: *A. crassna* is a monocotyledonous plant with simple leaves that vary from elliptical to obovate with smooth or undulating margins. The leaves are 2.5–5 centimetres wide and 7–11.5 centimetres long. Mature leaves are leathery, but young leaves are short and similar to tendrils. There are 12–18 pairs of leaf veins, and the leaf stalk is 3–7 millimetres long. The colour is light green, and mature leaves turn dark green before transitioning to yellow when they fall.

Flowers: They come in clusters at the leaf forks or the ends of branches, with 4–6 flowers. In the inflorescence, the stem is 3–5 mm, covered with soft hairs. The flowers have white or light-yellow petals. The perfect flower has a pedicel of 5–10 millimetres, covered with soft fur. The sepals are fused into a tubular shape, with five separated sepals measuring 2–3.5 millimetres wide and 3–4 millimetres long, with soft hairs on both outer and inner surfaces. The flower petals are white, with the base of the petals fused into a dome shape. There are long, soft hairs, measuring 1–1.5 millimetres, on the dome. The filament measures 1–1.5 millimetres, with a 1-millimetre anther. The ovary is elongated, measuring 2–3 millimetres. The style is not distinctly visible, and soft hairs cover the stigma, measuring 1 millimetre.

Fruits: When dried and mature, *A. crassna* fruits are nearly spherical, covered with soft hair, and measure about 2–2.5 centimetres in width and 2.5–3.5 centimetres in length. The fruits have persistent sepals that remain attached. The seeds are drop-shaped and measure 5 x 10 millimetres.

Habitat: *A. crassna* is found in upland areas in dry evergreen and tropical rainforests.

Distribution in Thailand: *A. crassna* can be found in the northern region (Phitsanulok, Phetchabun, Uttaradit), northeastern region (Loei), eastern region (Nakhon Ratchasima, Chaiyaphum, Nakhon Nayok), and southeast region (Prachin Buri, Sa Kaeo, Chanthaburi, Rayong, Trat).

International Distribution: *A. crassna* is also found in Laos, Cambodia, and Vietnam.

Conservation Status: According to the IUCN Red List (version 2.3, 2012), *A. crassna* is classified as critically endangered (CR A1cd).

3.1.2 Traditional Uses

Agarwood is also a fragrant wood that is culturally significant in many Asian countries, including China, Thailand, India, and Japan, as well as in various religious practices. Furthermore, Agarwood and other parts of the plant also play a role in traditional medicine in Southeast Asian countries, such as Chinese, Tibetan, Unani, and Ayurvedic medicines (Y. Z. H.-Y. Hashim et al., 2016; Swee & Chua, 2008). Agarwood is traded for various goods, including chips, flakes, powder, and wood products (Adhikari, Pokhrel, & Baral, 2021).

The oil is the most demanded feature of Agarwood. The oil extracted from Agarwood is widely used in cosmetics and personal care products, such as soap and shampoos. Notably, the oil functions as a fixative in agarwood perfumes made in both alcoholic and non-alcoholic carriers (Santoso, Irianto, Turjaman, & Sitepu, 2011; Sitepu et al., 2011). For example, Muslims use a perfume called "Attar" to lace prayer clothes. This water-based perfume contains Agarwood oil (Barden, Anak, Mulliken, & Song, 2000).

The solid pieces of Agarwood have been used for various purposes in many countries. While the bark of the Agarwood tree is used to make cloth and rope, the wood is widely utilised in India as a fuel for fumigation, joss sticks, and flea and louse repellents. In China, the wood of Agarwood is used for joss sticks and ornaments, but the bark is used to produce paper (Y. Z. H.-Y. Hashim et al., 2016). Several reviews also mention that in many places, solid shards of Agarwood can be crafted into bracelets, beads, and boxes (Barden et al., 2000; Persoon & van Beek, 2008).

3.2 PHYTOCHEMISTRY

A. crassna leaf contains bioactive substances such as mangiferin, iriflophenone 3-C-β-d-glucoside, and iriflophenone 3,5-C-β-d-diglucoside (E. Wongwad, Ingkaninan, Wisuitiprot, Sritularak, & Waranuch, 2020). The 2-(2-hydroxy-2-phenylethyl) chromones that were extracted from *A. crassna's* Agarwood showed antibacterial and inhibiting effects on acetylcholinesterase (AChE) (Wang et al., 2018). Sesquiterpenoids isolated from the ethyl acetate extract of Agarwood of *A. crassna* showed acetylcholinesterase inhibition activity and antibacterial activities (Wang et al., 2016). β-caryophyllene isolated from the essential oil of *A. crassna* demonstrated antioxidant, antibacterial, and antifungal activity. Several aquilacrassnins that are ester-bonded dimeric compounds and four crassins A-D that were bi-2-(2-phenylethyl) chromone derivatives were isolated from the ethyl acetate extract of Agarwood of *A. crassna* (Y. Yang et al., 2018). However, only crassins showed weak inhibitory activity against AChE and weak cytotoxicity against the human myeloid leukaemia cell line (K562) (Y. Yang et al., 2017; Zhao et al., 2021). Additionally, more than fifty chromones were detected in Agarwood samples from *A. crassna* (J. Yang et al., 2016).

3.3 MEDICINAL USES

3.3.1 Ethnopharmacology

Agarwood has been used in many different countries for medical purposes. The ethnopharmacological evidence suggests that Agarwood gained scientific interest due to its potential pharmacological activities, including antioxidant, anti-inflammatory, anti-diabetic, anticancer, and anti-depressant activities (S. S. Dahham et al., 2015; Tay, Tan, Abas, Yim, & Ho, 2014). Significantly, the use of Agarwood in anti-inflammatory treatment and related activities has been observed in many reviews, for example, to treat rheumatism, arthritis, body pain, asthma, and gout (Borris, Blaskó, & Cordell, 1988), use as cardioprotective agents (Owen & Johns, 2002), and to treat joint pain (Grosvenor, Gothard, McWilliam, Supriono, & Gray, 1995). Moreover, in some localities, the heartwood is interchangeably used with Agarwood to treat patients similarly (Y. Z. Hashim, P. G. Kerr, P. Abbas, & H. Mohd Salleh, 2016). Therefore, the use of Agarwood in terms of ethnopharmacology may differ in many countries or localities.

The plant is traditionally taken orally in Bangladesh to treat rheumatism (Parvez Rana, Shawkat Islam Sohel, Akhter, & Jahirul Islam, 2010). It has been widely used in traditional medicine in China to treat high fever, cough, rheumatism, and gastric symptoms (Commission, 2015). India has a rich tradition of utilising this plant for various ailments such as diarrhoea, dysentery, nausea, mouth and teeth diseases, trembling, sprains, bone fractures, asthma, anorexia, arthritis, and cardiac disorders (Warrier, 1993). In Indonesia, the ethnomedicinal use of the plant centres around the treatment of joint pain (Grosvenor et al., 1995). In Japan, Agarwood is a stomachic and sedative agent (Okugawa, Ueda, Matsumoto, Kawanishi, & Kato, 1993). Korean traditional medicine incorporates the plant for treating cough, acroparalysis, coup, and asthma, and as a stomachic agent, tonic, sedative, and expectorant (Bakansing, Sahri, & Liew, 2019; Y. Z. Hashim et al., 2016). Malaysian ethnobotanical practices involve using the plant as a tonic, stimulant, and carminative agent after childbirth and treating rheumatism and body pains (Y. Z. Hashim et al., 2016). In the Philippines, the plant is utilised to prevent wounds from bleeding (Lemmens & Bunyapraphatsara, 2003). In Thailand, the plant is recognised for its effectiveness in addressing various conditions, not only vomiting, nausea, and fainting (Suvitayavat, Tunlert, Thirawarapan, Kitpati, & Bunyapraphatsara, 2005) but also diarrhoea, dysentery, skin diseases, and as an antispasmodic and cardiovascular function enhancer

(Kamonwannasit et al., 2013). Finally, in Tibet, the plant is traditionally employed as a cardioprotective agent for treating nervous and emotional disorders (Owen & Johns, 2002). These ethnobotanical uses across different regions highlight the plant's diverse therapeutic potential and significance in traditional medicinal practices.

3.4 PHARMACOLOGICAL ACTIVITIES

Aquilaria crassna has been reported to exert various bioactivities, including anti-cancer, anti-inflammatory, antipyretic, antimicrobial, anti-ischaemic, laxative, and antioxidant properties. The plant materials of *A. crassna* contain sesquiterpenes, 2-(2-phenylethyl)-4H-chromen-4-one derivatives, genkwanins, mangiferins, iriflophenones, cucurbitacins, terpenoids, and phenolic acids, which contribute to its pharmacological activities (Adam, Lee, & Mohamed, 2017; Y. Yang et al., 2018). The literature review of the pharmacological activities of *A. crassna* is described below.

3.4.1 Antioxidant Activities

Both leaf and stem bark extracts exhibited potent antioxidant activities in various assays. These extracts demonstrated significant scavenging capabilities against free radicals, as indicated by low IC50 values in DPPH, ABTS, and FRAP assays in leaf extracts (Kamonwannasit et al., 2013; Pattarachotanant et al., 2022; Ray, Leelamanit, Sithisarn, & Jiratchariyakul, 2014; Eakkaluk Wongwad et al., 2019) and stem bark extract (Saad Sabbar Dahham et al., 2014).

3.4.2 Anticancer Activities

The plant's stem bark oil and ethanol extracts exhibit significant anticancer properties. In vitro studies demonstrated cell proliferation, migration, and colony formation inhibition, suggesting potential anti-metastatic activities. The extracts showed promising IC50 values, indicating their effectiveness against cancer cells (Saad Sabbar Dahham et al., 2014; S. S. Dahham et al., 2016; S. S. Dahham et al., 2015).

3.4.3 Anti-Inflammatory/Antipyretic Activities

The leaf's methanol extract and the heartwood's ethyl acetate extract showcase diverse pharmacological activities. *In vivo*, the leaf extract demonstrated antipyretic effects (Sattayasai, Bantadkit, Aromdee, Lattmann, & Airarat, 2012) and anti-inflammatory (Eakkaluk Wongwad et al., 2019), while the heartwood extract displayed inhibition of TNF-α production *in vitro* (Kumphune et al., 2011).

3.4.4 Antimicrobial Activities

The plant exhibits potent antimicrobial activities across various parts. Leaf extracts demonstrated inhibitory effects on bacterial growth and biofilm formation

(Kamonwannasit et al., 2013), while wood extracts displayed antifungal properties (Novriyanti, Santosa, Syafii, Turjaman, & Sitepu, 2010; Nun-Anan et al., 2021).

3.4.5 Anti-Ischaemic Activities

The heartwood's ethyl acetate extract exhibits promising anti-ischaemic properties, protecting cardiac cells *in vitro* and *ex vivo* from simulated ischaemic damage. The extract demonstrated increased cell viability, decreased LDH activity, and inhibition of p38MAPK phosphorylation (Jermsri & Kumphune, 2012; Suwannasing, Paiyabhroma, & Kumphune, 2012).

3.4.6 Laxative Effect

The plant's leaf extracts demonstrated a notable laxative effect *in vivo*, as observed in a study by Kakino et al. (2012). Administration of aqueous and ethanol extracts resulted in decreased intestinal transit time, with both extracts accelerating carmine egestion (Kakino et al., 2012).

3.4.7 Anti-ageing Effects

Intriguingly, research studies also demonstrated the anti-ageing potential of *A. crassna* leaves. Both ethanol and hexane extracts of *A. crassna* leaf were shown to extend the lifespan by preventing the accumulation of ROS and improving body size and length in *Caenorhabditis elegans*, a nematode. The lifespan-extending effects of hexane extracts were mainly modulated through the activation of genes within the DAF-16/FoxO pathway (Pattarachotanant et al., 2022).

3.4.8 Neuroprotective Effects

Both *in vitro* and *in vivo* studies have reported the neuroprotective effects of *A. crassna* leaves. For instance, the ethanolic extract from the leaves showed neuroprotective effects against glucose-induced neurotoxicity *in vitro* using SH-SY5Y cells (Pattarachotanant et al., 2022). Phenolic glycosides extracted from *A. crassna* leaves have been identified as potent neurogenic and neuroprotective agents (Supasuteekul et al., 2017).

3.4.9 Analgesic Activity

A. crassna aqueous extracts exhibited analgesic activity *in vivo* in mice using a hot plate test, a simple behavioural test used to screen the effectiveness of the analgesic. The hot plate test is a rapid, inexpensive, and reliable test that involves placing a plexiglass container holding mice on a hot plate that is thermostatically controlled at 50°C. The reaction time was defined as the time each mouse spent on the hot plate (until it licked or jumped in response to pain) with a 30-second cut-off period to prevent tissue damage. Mice were orally given either water or *A. crassna* aqueous extract, and the reaction time of each animal was determined at 30, 60, and

90 minutes after treatment. Mice fed with aspirin (150 mg/kg) served as positive control. The results showed that treatment with *A. crassna* aqueous extract substantially improved the mice's thermal threshold compared to those treated with water. The analgesic effect of aspirin became apparent 30, 60, and 90 minutes later. In contrast, 60 and 90 minutes after oral administration of the test extract, the effect of the aqueous extract (800 mg/kg) was reported. However, the magnitude of the analgesic effect of *A. crassna* was identical only after 90 minutes of oral administration (Sattayasai et al., 2012).

3.4.10 Cardiovascular Health Benefits

A. crassna leaf aqueous extract was also reported to exhibit a vasorelaxant effect and did not cause any cytotoxicity to vascular smooth muscle cells (Wisutthathum et al., 2019). Aqueous extract of *A. crassna* leaves and its significant constituent, mangiferin, induced vasodilation in the mesenteric artery. This effect was partially mediated by endothelium-dependent NO release and, at higher concentrations of the extract, by directly acting on vascular smooth muscle cells through the blocking of L-type voltage-operated Ca^{2+} channels. This suggests that *A. crassna* leaf extracts may help treat human cardiovascular dysfunction (Wisutthathum et al., 2019).

REFERENCES

Adam, A. Z., Lee, S. Y., & Mohamed, R. (2017). Pharmacological properties of agarwood tea derived from Aquilaria (Thymelaeaceae) leaves: An emerging contemporary herbal drink. *Journal of Herbal Medicine, 10*, 37–44. https://doi.org/10.1016/j.hermed.2017.06.002

Adhikari, S. R., Pokhrel, K., & Baral, S. D. (2021). Economic value of Agarwood and its prospects of cultivation. *International Journal of Applied Sciences and Biotechnology, 9*(1), 23–31.

Bakansing, S. M., Sahri, M. H., & Liew, K. C. (2019). Uses of agarwood. *Prospects and Utilisation of Tropical Plantation Trees*. Edited by Kang Chiang Liew. CRC Press: Boca Raton, FL, 179.

Barden, A., Anak, N. A., Mulliken, T., & Song, M. (2000). Heart of the matter: Agarwood use and trade and CITES implementation for Aquilaria malaccensis. *Traffic International*, Cambridge, UK, 17–18.

Borris, R. P., Blaskó, G., & Cordell, G. A. (1988). Ethnopharmacologic and phytochemical studies of the Thymelaeaceae. *Journal of Ethnopharmacology, 24*(1), 41–91. https://doi.org/10.1016/0378-8741(88)90138-9

Commission, C. P. (2015). Chinese pharmacopoeia. *China Medical Science Press: Beijing, China, 1*, 191–193.

Dahham, S. S., Ahamed, M. B. K., Saghir, S. M., Alsuede, F. S., Iqbal, M. A., & Majid, A. (2014). Bioactive essential oils from Aquilaria crassna for cancer prevention and treatment. *Global Journal of Pure and Applied Sciences, 4*, 26–31.

Dahham, S. S., Hassan, L. E., Ahamed, M. B., Majid, A. S., Majid, A. M., & Zulkepli, N. N. (2016). In vivo toxicity and antitumor activity of essential oils extract from Agarwood (Aquilaria crassna). *BMC Complementary and Alternative Medicine, 16*, 236. https://doi.org/10.1186/s12906-016-1210-1

Dahham, S. S., Tabana, Y. M., Iqbal, M. A., Ahamed, M. B., Ezzat, M. O., Majid, A. S., & Majid, A. M. (2015). The anticancer, antioxidant and antimicrobial properties of the sesquiterpene beta-caryophyllene from the essential oil of Aquilaria crassna. *Molecules, 20*(7), 11808–11829. https://doi.org/10.3390/molecules200711808

Forest Research and Development office, R. F. D., Ministry of Natural Resources and Environment. (2557). การวิจัยไม้กฤษณา.

Grosvenor, P. W., Gothard, P. K., McWilliam, N. C., Supriono, A., & Gray, D. O. (1995). Medicinal plants from Riau Province, Sumatra, Indonesia. Part 1: Uses. *Journal of Ethnopharmacology, 45*(2), 75–95. https://doi.org/10.1016/0378-8741(94)01209-I

Hashim, Y. Z. H.-Y., Kerr, P. G., Abbas, P., & Mohd Salleh, H. (2016). Aquilaria spp. (Agarwood) as source of health beneficial compounds: A review of traditional use, phytochemistry and pharmacology. *Journal of Ethnopharmacology, 189*, 331–360. https://doi.org/10.1016/j.jep.2016.06.055

Jermsri, P., & Kumphune, S. (2012). Ethylacetate extract of Aquilaria crassna preserve actin cytoskeleton on stimulated ischemia induced cardiac cell death. *Journal of Medicinal Plants Research, 6*(23), 4057–4062.

Kakino, M., Sugiyama, T., Kunieda, H., Tazawa, S., Maruyama, H., Tsuruma, K., . . . Mori, H. (2012). Agarwood (Aquilaria crassna) extracts decrease high-protein high-fat diet-induced intestinal putrefaction toxins in mice. *Pharm Anal Acta, 3*(152), 2.

Kamonwannasit, S., Nantapong, N., Kumkrai, P., Luecha, P., Kupittayanant, S., & Chudapongse, N. (2013). Antibacterial activity of Aquilaria crassna leaf extract against Staphylococcus epidermidis by disruption of cell wall. *Annals of Clinical Microbiology and Antimicrobials, 12*, 20. https://doi.org/10.1186/1476-0711-12-20

Kumphune, S., Prompunt, E., Phaebuaw, K., Sriudwong, P., Pankla, R., & Thongyoo, P. (2011). Anti-inflammatory effects of the ethyl acetate extract of Aquilaria crassna inhibits LPS-induced tumour necrosis factor-alpha production by attenuating P38 MAPK activation. *International Journal of Green Pharmacy, 5*(1), 43–48. https://doi.org/10.4103/0973-8258.82099

Lemmens, R. H. M. J., & Bunyapraphatsara, N. (2003). *Medicinal and poisonous plants 3*: Backhuys Publishers.

Novriyanti, E., Santosa, E., Syafii, W., Turjaman, M., & Sitepu, I. R. (2010). Anti fungal activity of wood extract of Aquilaria crassna Pierre ex Lecomte against agarwood-inducing fungi, Fusarium solani. *Indonesian Journal of Forestry Research, 7*(2), 155165.

Nun-Anan, P., Suchat, S., Mahathaninwong, N., Chueangchayaphan, N., Karrila, S., & Limhengha, S. (2021). Study of Aquilaria crassna wood as an antifungal additive to improve the properties of natural rubber as air-dried sheets. *Polymers (Basel), 13*(23). https://doi.org/10.3390/polym13234178

Okugawa, H., Ueda, R., Matsumoto, K., Kawanishi, K., & Kato, A. (1993). Effects of agarwood extracts on the central nervous system in mice. *Planta medica, 59*(1), 32–36.

Owen, P. L., & Johns, T. (2002). Antioxidants in medicines and spices as cardioprotective agents in Tibetan highlanders. *Pharmaceutical Biology, 40*(5), 346–357. https://doi.org/10.1076/phbi.40.5.346.8461

Parvez Rana, M., Shawkat Islam Sohel, M., Akhter, S., & Jahirul Islam, M. (2010). Ethno-medicinal plants use by the Manipuri tribal community in Bangladesh. *Journal of Forestry Research, 21*(1), 85–92. https://doi.org/10.1007/s11676-010-0015-1

Pattarachotanant, N., Sornkaew, N., Warayanon, W., Rangsinth, P., Sillapachaiyaporn, C., Vongthip, W., . . . Tencomnao, T. (2022). Aquilaria crassna leaf extract Ameliorates glucose-induced neurotoxicity in vitro and improves lifespan in Caenorhabditis elegans. *Nutrients, 14*(17). https://doi.org/10.3390/nu14173668

Persoon, G. A., & van Beek, H. H. (2008). Growing 'The Wood of The Gods': Agarwood production in Southeast Asia. In D. J. Snelder & R. D. Lasco (Eds.), *Smallholder Tree Growing for Rural Development and Environmental Services: Lessons from Asia* (pp. 245–262). Springer Netherlands.

The Plant List (2013). Retrieved from http://www.theplantlist.org/

Ray, G., Leelamanit, W., Sithisarn, P., & Jiratchariyakul, W. (2014). Antioxidative compounds from Aquilaria crassna leaves. *Mahidol University Journal of Pharmaceutical Sciences, 41*(4), 54–58. https://www.scopus.com/inward/record.uri?eid=2-s2.0-84994612019&partnerID=40&md5=733f2fd65721862d32dc09965506abef

Santoso, E., Irianto, R. S. B., Turjaman, M., & Sitepu, I. R. (2011). *Gaharu-Producing Tree Induction Technology.* Paper presented at the Proceeding of Gaharu Workshop, Development of Gaharu Production Technology.

Sattayasai, J., Bantadkit, J., Aromdee, C., Lattmann, E., & Airarat, W. (2012). Antipyretic, analgesic and anti-oxidative activities of Aquilaria crassna leaves extract in rodents. *J Ayurveda Integr Med, 3*(4), 175179. https://doi.org/10.4103/0975-9476.104427

Sitepu, I. R., Santoso, E., Siran, S. A., Turjaman, M. (2011). *Fragrant Wood Gaharu: When the Wild Can No Longer Provide*: Ministry of forestry of indonesia in cooperation with international tropical timber organisation.

Supasuteekul, C., Tadtong, S., Putalun, W., Tanaka, H., Likhitwitayawuid, K., Tengamnuay, P., & Sritularak, B. (2017). Neuritogenic and neuroprotective constituents from Aquilaria crassna leaves. *Journal of Food Biochemistry, 41*(3), e12365. https://doi.org/10.1111/jfbc.12365

Suvitayavat, W., Tunlert, S., Thirawarapan, S. S., Kitpati, C., & Bunyapraphatsara, N. (2005). Actions of Ya-hom, a herbal drug combination, on isolated rat aortic ring and atrial contractions. *Phytomedicine, 12*(8), 561–569. https://doi.org/10.1016/j.phymed.2004.03.014

Suwannasing, C., Paiyabhroma, N., & Kumphune, S. (2012). Anti-ischemic effect of ethyl acetate extract of Aquilaria crassna by attenuation of p38-MAPK activation. *Journal of Applied Pharmaceutical Science, 2*(10), 026–030.

Swee, L., & Chua, L. (2008). AGARWOOD (AQUILARIA MALACCENSIS) IN MALAYSIA.

Tay, P. Y., Tan, C. P., Abas, F., Yim, H. S., & Ho, C. W. (2014). Assessment of extraction parameters on antioxidant capacity, polyphenol content, epigallocatechin gallate (EGCG), epicatechin gallate (ECG) and iriflophenone 3-C-beta-glucoside of Agarwood (Aquilaria crassna) young leaves. *Molecules, 19*(8), 12304–12319. https://doi.org/10.3390/molecules190812304

Wang, H. N., Dong, W. H., Huang, S. Z., Li, W., Kong, F. D., Wang, H., . . . Dai, H. F. (2016). Three new sesquiterpenoids from Agarwood of Aquilaria crassna. *Fitoterapia, 114*, 7–11. https://doi.org/10.1016/j.fitote.2016.07.014

Wang, H. N., Mei, W. L., Dong, W. H., Kong, F. D., Li, W., Yuan, J. Z., & Dai, H. F. (2018). Two new 2-(2-Hydroxy-2-phenylethyl)chromens from Agarwood originating from Aquilaria crassna. *J Asian Nat Prod Res, 20*(2), 122–127. https://doi.org/10.1080/10286020.2017.1379998

Warrier, P. K. (1993). *Indian Medicinal Plants: A Compendium of 500 Species* (Vol. 5). Orient Blackswan.

Wisutthathum, S., Kamkaew, N., Inchan, A., Chatturong, U., Paracha, T. U., Ingkaninan, K., . . . Chootip, K. (2019). Extract of Aquilaria crassna leaves and mangiferin are vasodilators while showing no cytotoxicity. *J Tradit Complement Med, 9*(4), 237–242. https://doi.org/10.1016/j.jtcme.2018.09.002

Wongwad, E., Ingkaninan, K., Wisuitiprot, W., Sritularak, B., & Waranuch, N. (2020). Thermal degradation kinetics and pH-Rate profiles of Iriflophenone 3,5-C-β-d-diglucoside, Iriflophenone 3-C-β-d-Glucoside and mangiferin in Aquilaria crassna leaf extract. *Molecules, 25*(21). https://doi.org/10.3390/molecules25214898

Wongwad, E., Pingyod, C., Saesong, T., Waranuch, N., Wisuitiprot, W., Sritularak, B., . . . Ingkaninan, K. (2019). Assessment of the bioactive components, antioxidant, antiglycation and anti-inflammatory properties of Aquilaria crassna Pierre ex Lecomte leaves. *Industrial Crops and Products, 138*, 111448.

Yang, J., Dong, W., Kong, F., Liao, G., Wang, J., Li, W., . . . Dai, H. (2016). Characterisation and analysis of 2-(2-Phenylethyl)-chromone derivatives from Agarwood (Aquilaria crassna) by artificial holing for different times. *Molecules, 21*(7). https://doi.org/10.3390/molecules21070911

Yang, Y., Chen, H. Q., Kong, F. D., Zhou, L. M., Li, W., Dong, W. H., . . . Dai, H. F. (2018). Dimeric sesquiterpenoid-4H-chromone derivatives from Agarwood of Aquilaria crassna and their cytotoxicity. *Phytochemistry, 145*, 207–213. https://doi.org/10.1016/j.phytochem.2017.08.007

Yang, Y., Mei, W. L., Kong, F. D., Chen, H. Q., Li, W., Chen, Z. B., & Dai, H. F. (2017). Four new bi-2-(2-phenylethyl)chromone derivatives of Agarwood from Aquilaria crassna. *Fitoterapia, 119*, 20–25. https://doi.org/10.1016/j.fitote.2017.03.008

Zhao, Y. M., Yang, L., Kong, F. D., Dong, W. H., Li, W., Chen, H. Q., . . . Dai, H. F. (2021). Three new 5,6,7,8-tetrahydro-2-(2-phenylethyl)chromones and one new dimeric 2-(2-phenylethyl)chromone from Agarwood of Aquilaria crassna Pierre ex Lecomte in Laos. *Nat Prod Res, 35*(14), 2295–2302. https://doi.org/10.1080/14786419.2019.1672066

4 *Rhinacanthus nasutus* (L.) Kurz's Role in Cancer and Neurodegenerative Disease

Prevention and Treatment

James M. Brimson and Tewin Tencomano

4.1 INTRODUCTION TO *RHINACANTHUS NASUTUS*

Rhinacanthus nasutus (Figure 4.1) is a medicinal herb native to Thailand and found throughout South and Southeast Asia. It is known for its distinctive white/pale violet flower, from which it gets a number of its common names, such as snake jasmine and white crane flower. *R. nasutus* is known for its use as an antioxidant and for its anticancer properties in skin treatments.

4.1.1 Phytochemistry

There are a vast number of phytochemicals that can be isolated from *R. nasutus,* including a large group of naphthoquinones and rhinacanthins (Table 4.1). The rhinacanthins and naphthoquinones are considered the major specific active compounds. However, there are also many compounds less specific to *R. nasutus* that can be isolated in reasonable quantities. These include plant sterols such as lupeol, stigmasterol, beta-sitosterol, and beta-amyrin (Brimson et al. 2012), various flavonoids, benzenoids, quinones, and triterpenes (Huang et al. 2015). The extraction methods and techniques are covered in Chapter 2 of this book. *R. nasutus* is commonly extracted in either methanol or ethanol, and subsequent column separations (LC or HPLC) are used to separate specific compounds.

4.1.2 Cancer

R. nasutus extracts have been studied extensively concerning cancer (Brimson et al. 2020). The crude extract has been found to prevent tumour generation *in vivo,*

DOI: 10.1201/9781003508724-4

FIGURE 4.1 Photograph of *Rhinacanthus nasutus* showing the ovate leaves and white flowers.

with the extracts and several isolated compounds reducing the tumour size in mice with sarcoma cells implanted in them (Gotoh et al. 2004; Siripong et al. 2006a). Furthermore, a vast number of *in vitro* studies (Table 4.2) have identified the rhinacanthin compounds listed in Table 4.1 to prevent proliferation and induce apoptosis in cancer cell lines while having only having a minimal effect on non-tumourigenic cell lines (Kongkathip et al. 2004; Siripong et al. 2006a). Rhinacanthin C, N, and Q appear to have the most remarkable difference between cancer and non-cancer cell lines, suggesting that these compounds may be helpful as cancer treatments.

Rhinacanthin C, N, and Q can inhibit cell cycle progression, preventing cell proliferation by arresting the cells in the G2/M phase. All three of these compounds have also been shown to induce caspase-dependent apoptosis in a time- and dose-dependent manner in HeLa cells (Siripong et al. 2006b). Furthermore, Siripong et al. (2006b) show that genomic DNA is degraded, another hallmark of apoptosis, in response to Rhinacanthin N, as indicated by the TUNNEL assay and the smeared genomic DNA ladder.

Topoisomerase enzymes (1 and 2) make cuts in the DNA, remove a twist in the DNA helix before rejoining the DNA, and allow the tension in the DNA helix to be removed so that the DNA can be copied or transcribed. Topoisomerase enzymes are targets for cancer treatments, as inhibitors will prevent cell cycle progression, thereby preventing proliferation. Furthermore, the induction of apoptosis can be achieved depending on the type of inhibition; nicks and breaks in the DNA can be produced, which in turn triggers cell death.

Rhinacanthin derivatives such as N and Q can block the DNA relaxation brought about by Topoisomerase 2 (Kongkathip et al. 2013) by competitively blocking ATP binding to Topoisomerase 2 (Boonyalai et al. 2013) and inhibiting DNA binding itself (Kongkathip et al. 2013; Pradidphol et al. 2012). A structure-activity

TABLE 4.1
Structure and Use in Medicine of Rhinacanthin Molecules

Name	Structure	Medical Potential
Rhinacanthin A	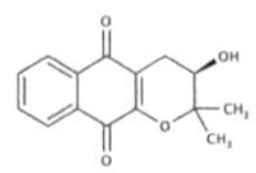	Cytotoxic activity
Rhinacanthin B	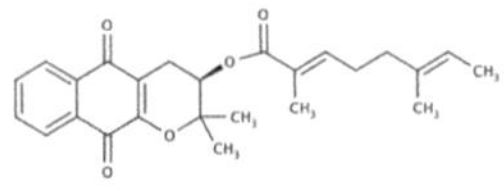	Inhibition against mosquito cytochrome P450 enzymes
Rhinacanthin C	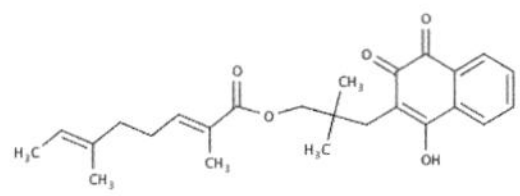	Anti-proliferative, antimicrobial, antiviral, inhibitor of P-gp and MRP2 efflux pumps, anti-allergic
Rhinacanthin D	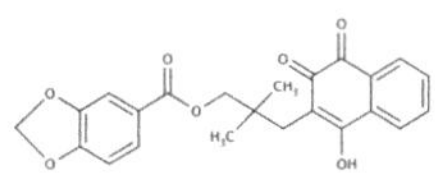	Inhibition of TNF α
Rhinacanthin E	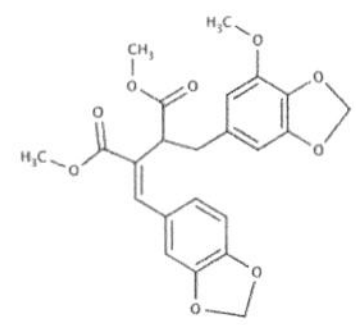	Inhibition of influenza virus
Rhinacanthin F	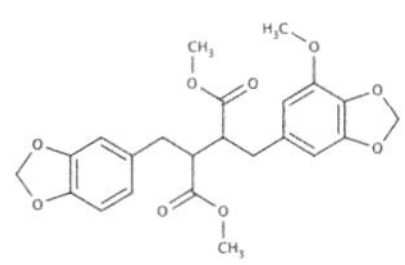	Inhibition of influenza virus
Rhinacanthin G	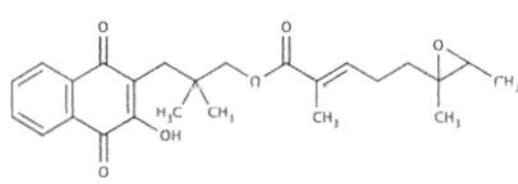	Anti-tumour activity
Rhinacanthin N	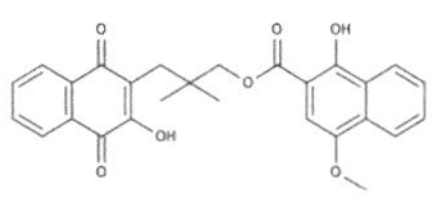	Anti-proliferative (HeLa53 tumours implanted in mice) anti-cancer, topoisomerase inhibitor
Rhinacanthin M	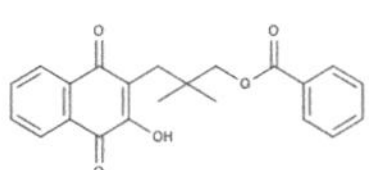	Anti-cancer, topoisomerase inhibitor
Rhinacanthin O	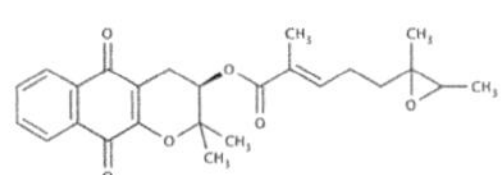	
Rhinacanthin Q	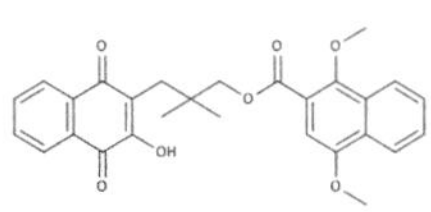	Anti-cancer, topoisomerase inhibitor
Rhinacanthone	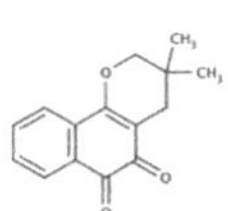	Anti-proliferative

TABLE 4.2
Compilation of IC50 Data for Various Cancer Cell Lines (KB, Hep-2, MCF7 HepG2, and HeLa) and One Non-tumourigenic Cell Line (Vero) Taken from Various Sources

	Antiproliferative Activity on Cell Line IC_{50}					
	KB	Hep-2	MCF7	HepG2	HeLa	Vero
μg/mL						
Root	4[a]	4[a]	20[a]	8.5[a]	0.4[a]	0.2[a]
Extract					1200[c]	
μM						
RN C	0.5[a]	0.8[a]	0.9[a]	0.4[a]	0.3[a] 26[c]	11[a]
RN D	0.5[a]	0.8[a]	15[a]	2[a]	0.5[a]	34[a]
RN G	5[a]	3[a]	9[a]	1[a]	5[a]	16[a]
RN O	6[a]	4[a]	8[a]	7[a]	6[a]	12[a]
RN M	0.3[a]	1[a]	3[a]	0.4[a]	0.9	36[a]
RN N	0.33[a] 0.2[b]	4[a]	3[a]	0.4[a, b]	0.8[a] 0.3[b]	13[a, b]
RNQ	4[a] 1.40[b]	3.6[a]	10.6[a]	1[a] 3[b]	1[a] 3.8[b]	41[a]
Rhinacanthone	4[a]	4[a]	5[a]	5[a]	4[a]	4[a]

[a] Data obtained from Siripong et al. (2006a).
[b] Data obtained from Kongkathip et al. (2004).
[c] Data obtained from Gotoh et al. (2004).

relationship exists with Rhinacanthin N (which contains an –OH group at the 2 position of the naphthalene moiety position see Figure 4.2) being a more potent inhibitor of Topoisomerase 2 and subsequently inducing cell death more than Rhinacanthin Q (which contains a -OMe group) or Rhinacanthin M (which includes no –OH group).

β-Sitosterol and other plant sterols (Figure 4.3) isolated from *R. nasutus* have been studied extensively for their antiproliferative effects on cancer cells. β-Sitosterol has been shown to inhibit the growth of HT-29 cells, altering membrane lipids and activating the sphingomyelin cycle (Awad et al. 1997; Awad et al. 1995); induce caspase, bax/bcl-2-dependent cell death in U937 leukemic cells (Park et al. 2007), and induce apoptosis in MDA-MB- 231 cells (Awad et al. 2000). However, other studies show β-Sitosterol can cause proliferation of oestrogen-responsive MCF7 breast cancer cells *in vivo*, with 150 μM β-Sitosterol resulting in a fivefold increase in the number of cells (Ju et al. 2004). The structures of β-Sitosterol (Figure 4.3) show a similarity. However, the affinity of β-Sitosterol for the oestrogen receptor is 100,000 times less than that of oestrogen (Kuiper et al. 1997). In the same study, Ju et al. (2004) show that in vitro MCF7 tumours implanted into mice were reduced in size when treated with β-Sitosterol. The study showed that β-Sitosterol reduced the blood levels of

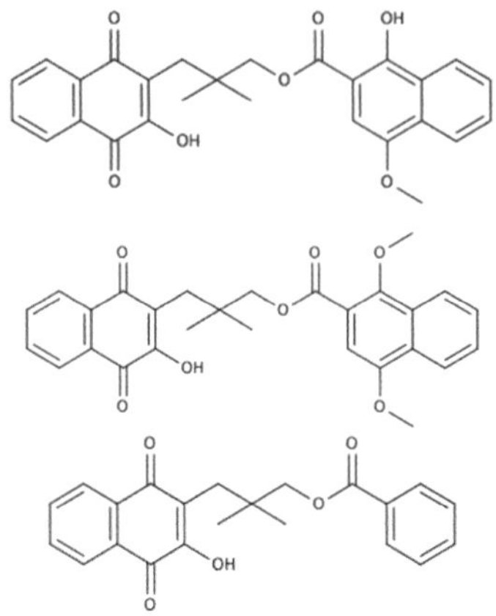

Rhinacanthin N

Rhinacanthin Q

Rhinacanthin M

FIGURE 4.2 Structures of Rhinacanthin N, Q, and M.

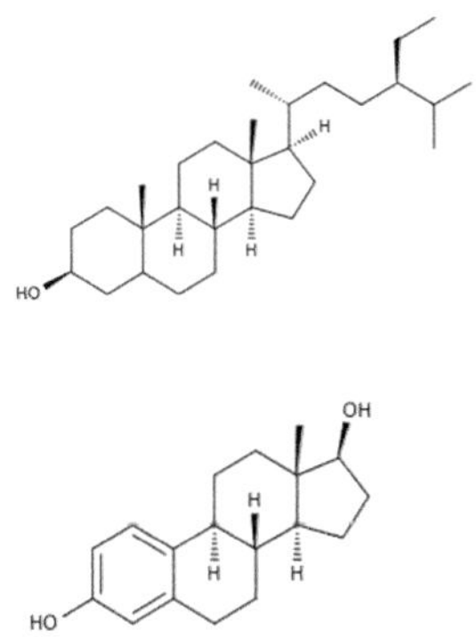

β sitosterol

Oestrogen

FIGURE 4.3 Structures of β sitosterol and oestrogen.

oestrogen, which is a more potent agonist of the oestrogen receptor, thus leading to a slower growth rate in the tumour.

4.1.3 Liver and Metabolic Diseases

The liver is a vital organ with several functions, including the detoxification of various metabolites, protein synthesis, and the production of biochemicals necessary for digestion. The liver is the first organ exposed to drugs, environmental toxins, and nutrients, which enter the liver via the hepatic portal vein of the digestive system. As such, the liver is vulnerable to injury due to chronic and acute toxicity.

R. nasutus extracts have been shown to protect the liver of rats exposed to aflatoxin B1 (Shyamal et al. 2010). Aflatoxin B1 causes hepatotoxicity by integrating with and damaging DNA and causing oxidative stress, which causes necrosis and can lead to liver cirrhosis and carcinoma of the liver. Aflatoxin B1 causes oxidative stress by depleting the cells' store of reduced glutathione (GSH), a natural cellular antioxidant, causing oxidative stress in the cell. Studies have shown that *R. nasutus* and lupeol can prevent the reduction in reduced glutathione, thus preventing oxidative stress (Preetha et al. 2006; Shyamal et al. 2010).

In Taiwan, *R. nasutus* is traditionally used to treat diabetes (Wu et al. 1988). It has been shown that leaf extracts of *R. nasutus*, at 200 mg/kg, can reduce the blood sugar of streptozotocin (a drug toxic to the insulin-producing beta cells of the pancreas) treated rats. In contrast, the same dose did not affect the blood sugar level of normal rats (Rao et al. 2012). Further research has shown that the induced diabetic rats have decreased levels of glucose-6-phosphate dehydrogenase (G6PDH), succinate dehydrogenase (SDH), and glutamate dehydrogenase (GDH), whereas there is an increase in lactate dehydrogenase (LDH). Treatment with 50 mg/kg/day of *R. nasutus* leaf extract for 30 days increased G6PDH, SDH, and GDH levels and reduced LDH levels (Rao et al. 2013). The hypoglycaemic effect of *R. nasutus* could come from molecules such as stigmasterol and β-sitosterol, which have been shown to reduce blood glucose levels synergistically (Jamaluddin et al. 1994). Another possible mechanism is that the rhinacanthin molecules activate the insulin receptor, as demonstrated by the 1,4-naphthoquinone group of molecules to which the rhinacanthins belong (He et al. 2011).

4.1.4 Neurological Diseases

The neuroprotective activity of *R. nasutus* has been shown in HT-22 hippocampus cells, where the pre-treatment of the cells with *R. nasutus* extracts from leaves or roots resulted in the protection of the cells against hypoxia-induced toxicity and glutamate-induced toxicity. Furthermore, subsequent studies showed that *R. nasutus* protected HT-22 cells from amyloid-β toxicity (Brimson et al. 2012; Brimson and Tencomnao 2014; Brimson and Tencomnao 2011). The protective effect of *R. nasutus* can be attributed to a number of compounds that have been identified in *R. nasutus* extracts, mainly acting through either anti-oxidative or anti-inflammatory mechanisms (Brimson and Tencomnao 2014; Brimson and Tencomnao 2013).

Deregulation of GSH in the brain is thought to be involved in several neurological disorders (Gawryluk et al. 2011). GSH is a potent, resident antioxidant found within cells, which helps neutralise free radical buildup because of ATP production. *R. nasutus* extracts have been shown to prevent the reduction in GSH levels brought about by various toxins (Shyamal et al. 2010). Furthermore, pure compounds isolated from *R. nasutus,* such as Lupeol (Figure 4.4), have been shown to have similar effects *in vivo* (Preetha et al. 2006).

Several studies have investigated the antidepressant effects of beta-amyrin. The studies show that 5–10 mg/kg injections of beta-amyrin can reduce the immobility time of rats subjected to the forced swim test (Subarnas et al. 1993c) (a behavioural

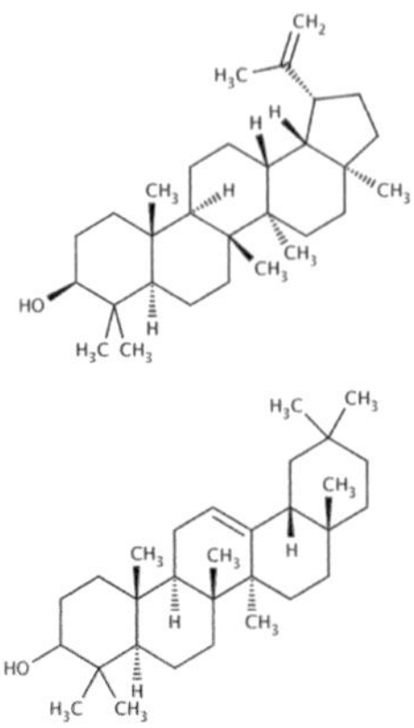

Lupeol

beta-amyrin

FIGURE 4.4 Structure of Lupeol and beta-amyrin.

paradigm that has been effective in the past at predicting antidepressant properties of drugs). Studies into the mechanism of action of beta-amyrin suggest that it potentially interacts with the GABA-a receptor (Aragão et al. 2006), the alpha-1-adrenoreceptors (Subarnas et al. 1993a), and causes the release of norepinephrine (Subarnas et al. 1993b).

In Alzheimer's disease, amyloid precursor protein (APP) is cut, releasing beta-amyloid protein. Beta-amyloid clumps together to form plaques, which in turn interfere with neurone cell signaling and induce an immune response in the brain, which leads to the destruction of disabled neurons. Furthermore, hyper-phosphorylation of Tau protein leads to neurofibrillary tangles, which block nutrients from getting to neurons and cause cell death. *R. nasutus* extracts have been shown to reduce cell death in cultured HT-22 cells that have been exposed to the toxic fragment of beta-amyloid protein (Brimson et al. 2012) and protect HT-22 cells from glucose deprivation (Brimson and Tencomnao 2011). Furthermore, *R. nasutus* can modulate immune responses, with Rhinacanthins C, N, and D inhibiting nitric oxide, TNF-α, and prostaglandin E2 in LPS-stimulated macrophage-like RAW 264.7 cells (Tewtrakul et al. 2009). This evidence suggests that *R. nasutus* has the potential for neurodegenerative disease treatment.

It has also been demonstrated that Rhinacanthin C reduces neuroinflammation induced by lipopolysaccharide (LPS), amyloid-β, and interferon-γ in BV-2 cells and primary rat glial cells and protects primary rat neurones from damage caused by β-amyloid toxicity (Chuang et al. 2017). Additionally, Rhinacanthin C has been shown to possess antioxidant properties and can inhibit the expression of interleukin-1β (IL-1β), interleukin-6 (IL-6), and tumour necrosis factor-alpha (TNF-α) mRNA in animals subjected to subarachnoid haemorrhage (Chang et al. 2016).

Rhinacanthin C and 2-hydroxy-1,4-naphthoquinone isolated from *R. nasutus* exhibit neuroprotective effects against glutamate-induced apoptosis in HT-22 cells. Additionally, Rhinacanthin C was found to inhibit autophagy and elevate ER stress.

However, Rhinacanthin C and 2-hydroxy-1,4-naphthoquinone mitigated ER stress and suppressed CHOP expression at low concentrations. Rhinacanthin C, at all tested concentrations, effectively prevented ROS production and ERK1/2 phosphorylation (Rakkhittawattana et al. 2022). Furthermore, a recent study in rats demonstrated that Rhinacanthin C exhibits anti-parkinsonian effects without impacting liver enzymes or other organs. The study also revealed that Rhinacanthin C could elevate monoamine levels in the brain (Saleem et al. 2021).

4.2 CONCLUSIONS

R. nasutus is a medicinal herb with wide-ranging applications within traditional medicine. *R. nasutus* is a dynamic medicinal herb that can potentially treat several diseases and ailments. The mechanisms by which this herb exerts its activities are beginning to be elucidated, with the isolated compounds showing promise in cancer research, neurodegenerative diseases, and liver and metabolic diseases.

REFERENCES

Aragão GF, Carneiro LM V, Junior a PF, et al. (2006) A possible mechanism for anxiolytic and antidepressant effects of alpha- and beta-amyrin from Protium heptaphyllum (Aubl.) March. *Pharmacol Biochem Behav,* 85: 827–834. doi: 10.1016/j.pbb.2006.11.019

Awad AB, Chen YC, Fink CS, Hennessey T (1995) beta-Sitosterol inhibits HT-29 human colon cancer cell growth and alters membrane lipids. *Anticancer Res,* 16: 2797–2804.

Awad AB, Downie AC, Fink CS (2000) Inhibition of growth and stimulation of apoptosis by beta-sitosterol treatment of MDA-MB-231 human breast cancer cells in culture. *Int J Mol Med,* 5: 541–546.

Awad AB, Von Holtz RL, Cone JP, et al. (1997) beta-Sitosterol inhibits growth of HT-29 human colon cancer cells by activating the sphingomyelin cycle. *Anticancer Res,* 18: 471–473.

Boonyalai N, Sittikul P, Pradidphol N, Kongkathip N (2013) Biophysical and molecular docking studies of naphthoquinone derivatives on the ATPase domain of human Topoisomerase II. *Biomed Pharmacother,* 67: 122–128. doi: 10.1016/j.biopha.2012.08.005

Brimson JM, Brimson SJ, Brimson CA, et al. (2012) Rhinacanthus nasutus extracts prevent glutamate and amyloid-β neurotoxicity in HT-22 mouse hippocampal cells: Possible active compounds include Lupeol, Stigmasterol and β-Sitosterol. *Int J Mol Sci,* 13: 5074–5097. doi: 10.3390/ijms13045074

Brimson JM, Tencomnao T (2011) Rhinacanthus nasutus protects cultured neuronal cells against hypoxia induced cell death. *Mol,* 16: 6322–6338.

Brimson JM, Tencomnao T (2013) Medicinal herbs and antioxidants: Potential of Rhinacanthus nasutus for disease treatment? *Phytochem Rev,* 13: 643–651. doi: 10.1007/s11101-013-9324-2

Brimson JM, Tencomnao T (2014) Rhinacanthus Nasutus extract as a neuroprotectant. In *Bioactive Nutracueticals and Dietary Supplaments in Neurological and Brain Disease.* Academic Press, Cambridge MA, USA. 2015. 77–84. doi: 10.1016/B978-0-12-411462-3.00009-6

Brimson JM, Prasanth MI, Malar DS, Brimson S, Tencomnao T (2020) Rhinacanthus nasutus "tea" infusions and the medicinal benefits of the constituent phytochemicals. *Nutrients,* 12(12): 3776.

Chang CZ, Wu SC, Kwan AL, Lin CL (2016) Rhinacanthin-C, a fat-soluble extract from Rhinacanthus nasutus, modulates high-mobility group box 1-related neuro-inflammation and subarachnoid hemorrhage-induced brain apoptosis in a rat model. *World Neurosurg*, 86: 349–360.

Chuang KA, Li MH, Lin NH, Chang CH, Lu IH, Pan IH, Takahashi T, Perng MD, Wen SF (2017) Rhinacanthin C Alleviates Amyloid- β Fibrils' toxicity on neurons and attenuates neuroinflammation triggered by LPS, Amyloid- β, and Interferon- γ in Glial cells. *Oxid Med Cell Longev,* 2017(1): 5414297.

Gawryluk JW, Wang J-F, Andreazza AC, et al. (2011) Decreased levels of glutathione, the major brain antioxidant, in post-mortem prefrontal cortex from patients with psychiatric disorders. *Int J Neuropsychopharmacol,* 14: 123–130.

Gotoh A, Sakaeda T, Kimura T, et al. (2004) Antiproliferative activity of Rhinacanthus nasutus (L.) Kurz extracts and the active moiety, Rhinacanthin C. *Biol Pharm Bull,* 27: 1070–1074.

He K, Chan C-B, Liu X, et al. (2011) Identification of a molecular activator for insulin receptor with potent anti-diabetic effects. *J Biol Chem*, 286: 37379–3788. doi: 10.1074/jbc.M111.247387

Huang RT, Lu YF, Inbaraj BS, Chen BH (2015) Determination of phenolic acids and flavonoids in Rhinacanthus nasutus (L.) kurz by high-performance-liquid-chromatography with photodiode-array detection and tandem mass spectrometry. *J Funct Foods,* 12: 498–508. doi: 10.1016/j.jff.2014.12.002

Jamaluddin F, Mohamed S, Lajis N (1994) Hypoglycaemic effect of Parkia speciosa seeds due to the synergistic action of –sitosterol and stigmasterol. *Food Chem*, 49: 339–345.

Ju YH, Clausen LM, Allred KF, et al. (2004) beta-Sitosterol, beta-Sitosterol glucoside, and a mixture of beta-Sitosterol and beta-Sitosterol glucoside modulate the growth of estrogen-responsive breast cancer cells in vitro and in ovariectomized athymic mice. *J Nutr,* 134: 1145–1151.

Kongkathip B, Akkarasamiyo S, Hasitapan K, et al. (2013) Synthesis of novel naphthoquinone aliphatic amides and esters and their anticancer evaluation. *Eur J Med Chem,* 60: 271–284. doi: 10.1016/j.ejmech.2012.12.006

Kongkathip N, Luangkamin S, Kongkathip B, et al. (2004) Synthesis of novel rhinacanthins and related anticancer naphthoquinone esters. *J Med Chem*, 47: 4427–4438. doi: 10.1021/jm030323g

Kuiper GG, Carlsson B, Grandien K, et al. (1997) Comparison of the ligand binding specificity and transcript tissue distribution of estrogen receptors alpha and beta. *Endocrinology,* 138: 863–870. doi: 10.1210/endo.138.3.4979

Park C, Moon D-O, Rhu C-H, et al. (2007) Beta-sitosterol induces anti-proliferation and apoptosis in human leukemic U937 cells through activation of caspase-3 and induction of Bax/Bcl-2 ratio. *Biol Pharm Bull,* 30: 1317–1323. doi: 10.1248/bpb.30.1317

Pradidphol N, Kongkathip N, Sittikul P, et al. (2012) First synthesis and anticancer activity of novel naphthoquinone amides. *Eur J Med Chem,* 49: 253–270. doi: 10.1016/j.ejmech.2012.01.020

Preetha SP, Kanniappan M, Selvakumar E, et al. (2006) Lupeol ameliorates aflatoxin B1-induced peroxidative hepatic damage in rats. *Comp Biochem Physiol Part C Toxicol Pharmacol,* 143: 333–339. doi: 10.1016/j.cbpc.2006.03.008

Rakkhittawattana V, Panichayupakaranant P, Prasanth MI, Brimson JM, Tencomnao T (2022) Rhinacanthin-C but not-D extracted from Rhinacanthus nasutus (L.) Kurz offers neuroprotection via ERK, CHOP, and LC3B pathways. *Pharmaceuticals,* 15(5): 627.

Rao P, Madhavi K, Dhananjaya Naidu M, Gan SH (2013) Rhinacanthus nasutus Ameliorates cytosolic and mitochondrial enzyme levels in streptozotocin-induced diabetic rats. *Evid Based Complement Alternat Med,* 2013: 486047. doi: 10.1155/2013/486047

Rao PV, Sujana P, Vijayakanth T, Naidu MD (2012) Rhinacanthus nasutus – Its protective role in oxidative stress and antioxidant status in streptozotocin induced diabetic rats. *Asian Pacific J Trop Dis,* 2: 327–330. doi: 10.1016/S2222-1808(12)60071-1

Saleem U, Gull Z, Saleem A, Shah MA, Akhtar MF, Anwar F, Ahmad B, Panichayupakaranant P (2021) Appraisal of anti-Parkinson activity of rhinacanthin-C in haloperidol-induced parkinsonism in mice: A mechanistic approach. *J Food Biochem,* 45(4), e13677.

Shyamal S, Latha PG, Suja SR, et al. (2010) Hepatoprotective effect of three herbal extracts on aflatoxin B1-intoxicated rat liver. *Singapore Med J,* 51: 326–331.

Siripong P, Kanokmedakul K, Piyaviriyagul S, et al. (2006a) Antiproliferative naphthoquinone esters from Rhinacanthus nasutus Kurz. roots on various cancer cells. *J Tradit Med,* 23: 166–172.

Siripong P, Yahuafai J, Shimizu K (2006b) Induction of apoptosis in tumor cells by three naphthoquinone esters isolated from Thai medicinal plant: Rhinacanthus nasutus KURZ. *Biol Pharm Bull,* 29: 2070–2076.

Subarnas A, Tadano T, Kisara K, Ohizumi Y (1993a) An alpha-adrenoceptor-mediated mechanism of hypoactivity induced by beta-amyrin palmitate. *J Pharm Pharmacol,* 45: 1006–1008.

Subarnas A, Tadano T, Nakahata N, et al. (1993b) A possible mechanism of antidepressant activity of beta-amyrin palmitate isolated from Lobelia inflata leaves in the forced swimming test. *Life Sci,* 52: 289–296.

Subarnas A, Tadano T, Oshima Y, et al. (1993c) Pharmacological properties of beta-amyrin palmitate, a novel centrally acting compound, isolated from Lobelia inflata leaves. *J Pharm Pharmacol,* 45: 545–550.

Tewtrakul S, Tansakul P, Panichayupakaranant P (2009) Effects of rhinacanthins from Rhinacanthus nasutus on nitric oxide, prostaglandin E2 and tumor necrosis factor-alpha releases using RAW264.7 macrophage cells. *Phytomedicine*, 16: 581–585. doi: 10.1016/j.phymed.2008.12.022

Wu TS, Tien H-J, Yeh M-Y, Lee K-H (1988) Isolation and cytotoxicity of rhinacanthin-A and -B, Two naphthoquinones from Rhinacanthus Nasutus. *Phytochemistry*, 27: 3787–3788.

5 The Role of *Clerodendrum petasites* S. Moore in Disease Treatment

Premrutai Thitilertdecha, James M. Brimson, and Nattawat Onlamoon

5.1 INTRODUCTION: *CLERODENDRUM PETASITES* S. MOORE

Clerodendrum petasites S. Moore (Figure 5.1), commonly known as One Root Plant or Thao-Yaai-Mom in Thai, belongs to the Lamiaceae family, as indicated by the International Plant Names Index (IPNI) (*Clerodendrum petasites* S. Moore, 2005). This plant is extensively found in various regions of Thailand, including the middle, northeastern, and southern parts. Traditional Thai practitioners have incorporated the plant into ethno-medical practices, utilising it to address conditions such as asthma (Hazekamp et al., 2001; Panthong et al., 2003), inflammation (Panthong et al., 1986), fever, cough, and vomiting (Panthong et al., 2003; Thai traditional medical textbook: Paet-Ta-Ya-Saat-Song-Kror, 2007). Additionally, the plant is traditionally employed in poultices for skin disorders like rash, abscess, urticaria, snake bites, and insect bites (Panthong et al., 2003; Thai traditional medical textbook: Paet-Ta-Ya-Saat-Song-Kror, 2007).

5.2 PHYTOCHEMISTRY

While the chemical components of the *Clerodendrum* genus have been extensively studied, there is limited research on *C. petasites*. Table 5.1 presents the identified compounds along with their chemical structures in the aerial parts, roots, and whole plant of *C. petasites*, drawing from studies by Hazekamp et al. (2001), Klaiklay (2009), Singharachai et al. (2011), Thitilertdecha et al. (2014a), and Thongchai et al. (2007).

 DOI: 10.1201/9781003508724-5

FIGURE 5.1 Photograph of *Clerodendrum petasites* S. Moore. Image kindly provided by Dr. Onusa Thamsermang, Chiang Mai University, Thailand.

5.3 MEDICAL EFFECTS OF PLANT EXTRACTS AND CHEMICAL CONSTITUENTS

5.3.1 Inflammation

While vanillic acid has been extensively studied for various bioactivities, its precise mechanism of action remains unclear. Studies suggest that vanillic acid can inhibit cyclooxygenase-2 (COX-2) (Figure 5.2), leading to a subsequent reduction in prostaglandin E2 (PGE2) production (Kim et al., 2011; Kim et al., 2010). Additionally, vanillic acid has been reported to dephosphorylate inhibitory kappa-B (IκB), diminishing the nuclear translocation of p65 and thereby reducing nuclear factor kappa-B (NF-κB) activity, along with subsequent proinflammatory cytokines such as tumour necrosis factor (TNF)-α and interleukin (IL)-6 (Kim et al., 2011, Kim et al., 2010). The receptor-interacting protein (RIP)-2/caspase-1 pathway and nitric oxide (NO) production are also deactivated by vanillic acid, contributing to its anti-inflammatory response (Kim et al., 2011). In contrast, the anti-inflammatory properties of nepetin have been solely reported through NF-κB deactivation (Clavin et al., 2007).

Verbascoside exhibits anti-inflammatory effects by inducing Src homology region 2 domain-containing phosphatase-1 (SHP-1), a negative regulator of downstream signalling pathways, such as transforming growth factor b-activated kinase-1 (TAK1). Consequently, TAK1 and c-Jun N-terminal kinase (JNK) are dephosphorylated, and nuclear translocation of activator protein 1 (AP-1) is attenuated. This suppresses COX-2 and inducible nitric oxide synthase (iNOS) activities through the SHP-1/TAK1/JNK/AP-1 pathway (Pesce et al., 2015). Supporting its anti-inflammatory properties, Jing et al. (2015) demonstrated reduced levels of proinflammatory cytokines, including TNF-α, interleukin (IL)-1β, and IL-6, attributed to verbascoside's

TABLE 5.1
Natural Bioactive Compounds Identified in *C. petasites*

Name	Structure	Examples of Potential Uses in Medicine
Vanillic acid	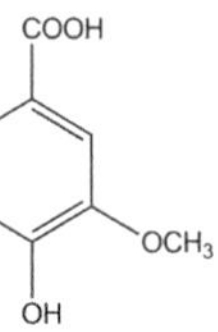	Pain (Yrbas et al., 2015), inflammatory diseases (Kim et al., 2011, Kim et al., 2010), neurodegenerative disorders (Singh et al., 2015)
4-Coumaric acid	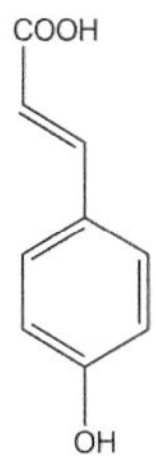	Infectious diseases (Chiang et al., 2003) and ischaemia (Guven et al., 2015)
Ferulic acid	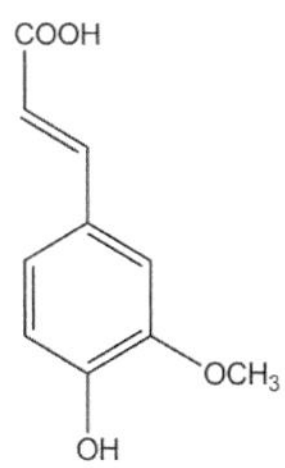	Infectious diseases (Chiang et al., 2003) and skin cancer (Ambothi et al., 2015)
Nepetin (6-methoxyluteolin)	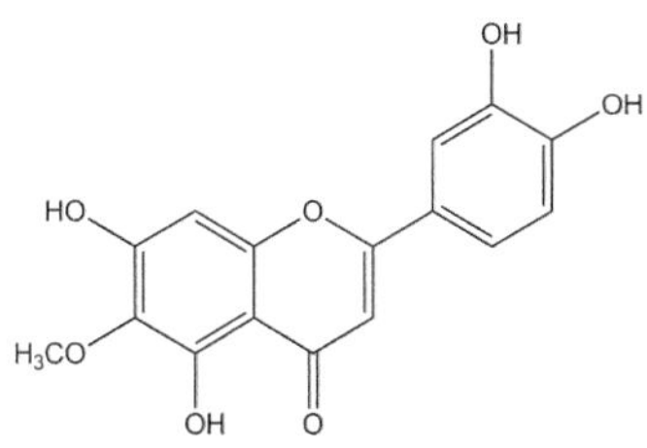	Bacterial infection (Sultana and Afolayan, 2007) and inflammatory diseases (Clavin et al., 2007)
Luteolin	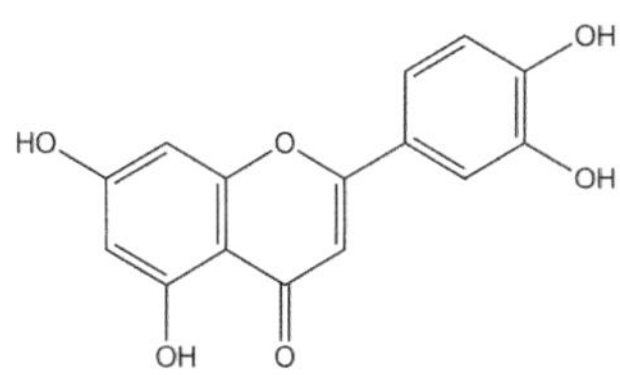	Liver cancer (Rafacho et al., 2015) and obesity (Kwon et al., 2015)
Apigenin	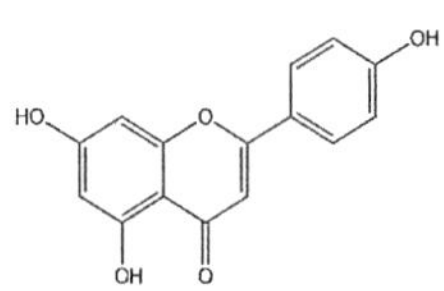	Epilepsy (Ying Chang et al., 2015) and colorectal cancer (Banerjee and Mandal, 2015)

(Continued)

TABLE 5.1 CONTINUED
Natural Bioactive Compounds Identified in *C. petasites*

Name	Structure	Examples of Potential Uses in Medicine
Naringenin	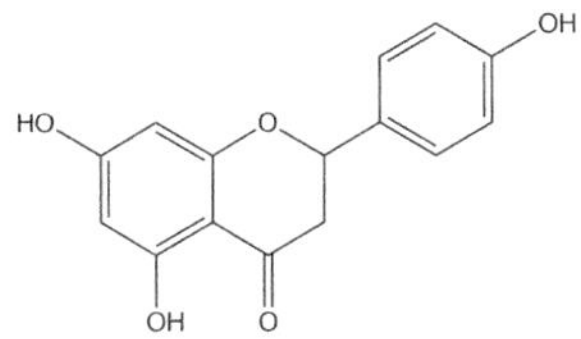	Cancer (Islas et al., 2015, Yen et al., 2015)
Hispidulin	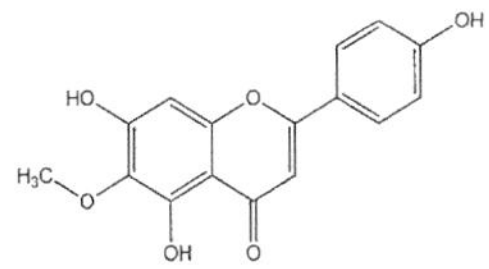	Inflammatory diseases (Gil et al., 1994), hyper-dopaminergic disorders (Huang et al., 2015), cancers (Gao et al., 2014), osteolytic bone diseases (Nepal et al., 2013) , and epilepsy (Kavvadias et al., 2004)
Hesperetin	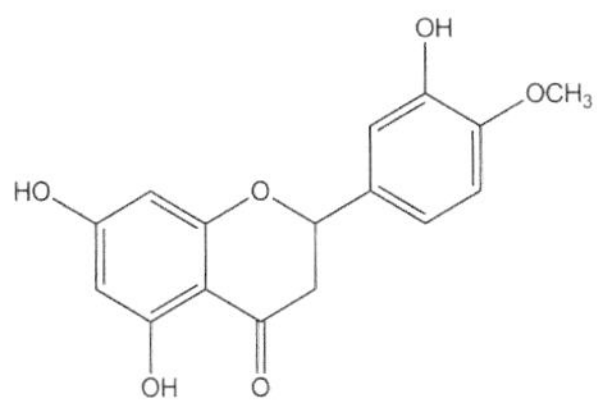	Diabetic retinopathy (Kumar et al., 2013)
Chrysin	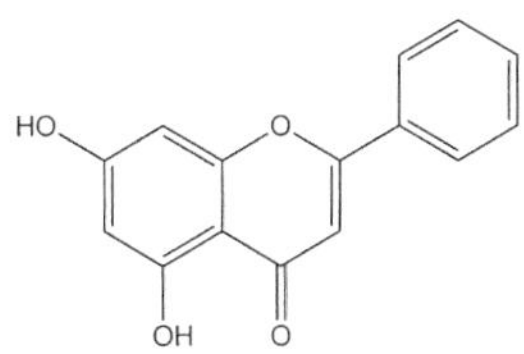	Airway inflammation (Shen et al., 2015)
Arbutin	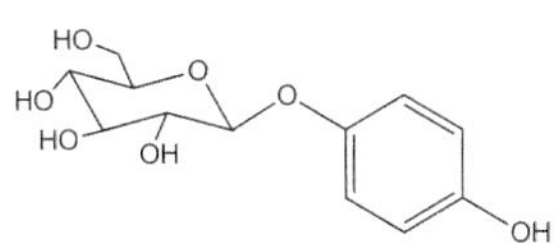	Gastric ulcer (Taha et al., 2012)
Verbascoside (acteoside)	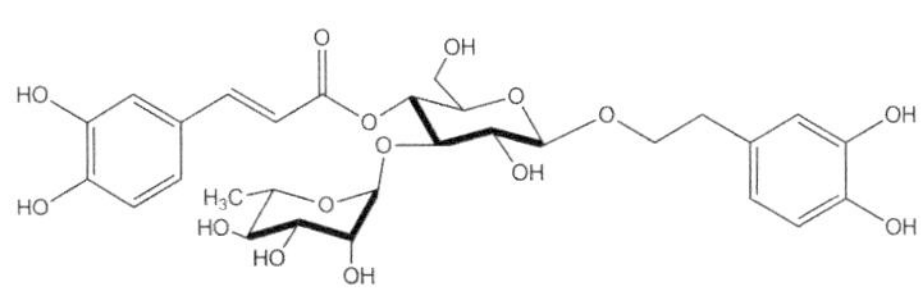	Bacterial infection (Shikanga et al., 2010), cardiovascular (Campo et al., 2015), inflammatory (Jing et al., 2015, Pesce et al., 2015), Alzheimer's diseases (Gao et al., 2015b, Peng et al., 2015), and cancer (Zhou et al., 2014a)

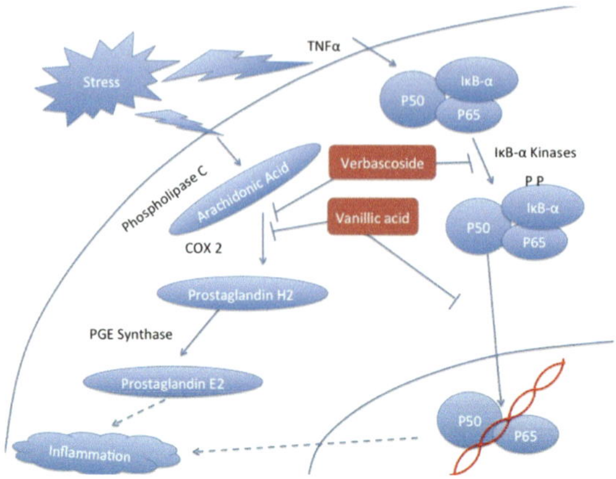

FIGURE 5.2 The anti-inflammatory effects of verbacoside and vanillic acid.

inhibitory effect on the phosphorylation of IκBα, NF-κB, p65, inhibitor of nuclear factor kappa-B kinase (IKK)-α, and IKK-β.

5.3.2 Cancer

The biological activities of hispidulin have been extensively investigated, with many studies focusing on hypoxia-inducible factor 1 alpha (HIF-1α), which plays an essential role in angiogenesis, cell proliferation, apoptosis, epithelial-mesenchymal transition (EMT), and invasiveness of cancer cells. Hispidulin is reported to inhibit HIF-1α function by modulating several upstream signalling pathways. Lin et al. (2010) demonstrated that hispidulin-induced AMP-activated protein kinase (AMPK) results in a suppression of eukaryotic initiation factor 4E-binding protein-1 (4E-BP1) through the mammalian target of rapamycin (mTOR) pathway and, therefore, a reduction of HIF-1α. This decreased the proliferation of cancer cells and induced apoptosis. Moreover, it was shown that lipogenic enzymes, including fatty acid synthase (FASN) and acetyl-CoA carboxylases (ACCs), were decreased after the AMPK stimulation, causing a reduction in fatty acid synthesis and, again, reducing the proliferation of cancer cells.

Besides the AMPK pathway, hispidulin was found to modulate the phosphatase and tensin homolog (PTEN)/phosphoinositide 3-kinase (PI3k)/Akt pathway, resulting in the inhibition of HIF-1α (Xie et al., 2015). He et al. (2011) showed that hispidulin suppressed the phosphorylation of vascular endothelial growth factor receptor 2 (VEGFR2), leading to PI3k/Akt/mTOR blockage. A deactivation of ribosomal protein S6 kinase (S6K) was also caused by mTOR suppression, resulting in the downregulation of tumour angiogenesis. Furthermore, hispidulin treatment–induced myeloid leukaemia cell differentiation protein Mcl-1, an antiapoptotic protein (Yang et al., 2010). The Mcl-1 downregulation leads to a massive decrease in protein translation, inducing tumour cell apoptosis.

Hispidulin interacts with tumour necrosis factor (TNF)-related apoptosis-inducing ligand (TRAIL) cell death pathway. Caspases-3 and -8 are activated by hispidulin, enhancing TRAIL-induced apoptosis (Yang et al., 2010). Caspase-3 was also consequentially activated by hispidulin-induced excessive reactive oxygen species (ROS) generation (Gao et al., 2014). Hispidulin has also been shown to interfere with cell cycle regulation (Gao et al., 2014; Lin et al., 2010), inhibiting cyclin D1 activity, resulting in the accumulation of G0/G1 phase and not progressing to S and G2/M phases. In terms of chemotherapy, hispidulin was reported to reduce the chemoresistance of cancer cells by downregulation of P-glycoproteins (P-gp) (Gao et al., 2015a).

5.3.3 Osteolytic Bone Diseases

Hispidulin has been extensively studied for its biological activities, with a particular focus on hypoxia-inducible factor 1 alpha (HIF-1α), a pivotal factor in angiogenesis, cell proliferation, apoptosis, epithelial-mesenchymal transition (EMT), and cancer cell invasiveness. The inhibitory effects of hispidulin on HIF-1α involve the modulation of various upstream signalling pathways. Lin et al. (2010) demonstrated that hispidulin activates AMP-activated protein kinase (AMPK), leading to the suppression of eukaryotic initiation factor 4E-binding protein-1 (4E-BP1) through the mammalian target of rapamycin (mTOR) pathway, ultimately reducing HIF-1α, cell proliferation, and inducing apoptosis. Additionally, hispidulin was found to impact the phosphatase and tensin homolog (PTEN)/phosphoinositide 3-kinase (PI3K)/Akt pathway, inhibiting HIF-1α (Xie et al., 2015). He et al. (2011) demonstrated hispidulin's suppression of vascular endothelial growth factor receptor 2 (VEGFR2) phosphorylation, blocking the PI3K/Akt/mTOR pathway and downregulating tumour angiogenesis. Hispidulin treatment also induced the anti-apoptotic protein myeloid leukaemia cell differentiation protein Mcl-1 (Yang et al., 2010), causing a significant decrease in protein translation and tumour cell apoptosis.

Furthermore, hispidulin interacts with the tumour necrosis factor (TNF)-related apoptosis-inducing ligand (TRAIL) cell death pathway, activating caspases-3 and -8 and enhancing TRAIL-induced apoptosis (Yang et al., 2010). Excessive reactive oxygen species (ROS) generation by hispidulin also activates caspase-3 (Gao et al., 2014). Hispidulin disrupts cell cycle regulation (Gao et al., 2014; Lin et al., 2010), inhibiting cyclin D1 activity and causing G0/G1 phase accumulation, preventing progression to the S and G2/M phases. Regarding chemotherapy, hispidulin was reported to reduce chemoresistance in cancer cells by downregulating P-glycoproteins (P-gp) (Gao et al., 2015a).

5.3.4 Neurological Diseases

In the brain, glutamate is a crucial excitatory neurotransmitter involved in memory, learning, and synaptic plasticity. However, an excess of glutamate can lead to toxicity, contributing to the pathophysiology of various diseases such as Alzheimer's disease, Parkinson's disease, ischaemia, and stroke. Apigenin has been noted for its ability to prevent brain damage induced by ischaemia, promote neuron survival, and ameliorate cognitive impairment in animal studies. One potential mechanism through which apigenin may mitigate neuron damage in these conditions is by preventing the excessive release of glutamate (Ying Chang et al., 2015).

A substantial increase in glutamate is strongly associated with epilepsy, making the moderation of glutamatergic neurotransmission a target for treatment. Lin et al. (2012) reported that hispidulin reduced the phosphorylation of extracellular signal-regulated kinase 1 and 2 (ERK 1/2) and the phosphorylation of vesicle-associated protein synapsin I, leading to a decrease in glutamate levels. Hispidulin also inhibited presynaptic voltage-dependent calcium entry, specifically targeting Cav2.2 and Cav2.1 channels. Kavvadias et al. (2004) demonstrated that hispidulin penetrated the blood-brain barrier, exerting an anticonvulsive effect by modulating gamma-aminobutyric acid (GABA) receptors (GABAA receptors) and enhancing GABA-induced chloride currents.

Alzheimer's disease is, in part, a result of oxidative stress causing damage to brain cells. It has been shown that verbascoside attenuates nitric oxide (NO) levels by reducing NO synthase (NOS) and deactivates Caspase-3, a pro-apoptotic protein that initiates neuronal apoptosis. Another study by Gao et al. (2015b) demonstrated that verbascoside enhances the expression of nerve growth factor (NGF) and tropomyosin receptor kinase A (TrkA), which are crucial factors for the survival of basal forebrain cholinergic neurons (BFCN).

5.4 CONCLUSIONS

Drawing insights from the documented properties of four primary active compounds – namely, vanillic acid, nepetin, hispidulin, and verbascoside – it is discerned that *C. petasites* stands out as a promising candidate for conferring anti-inflammatory, anti-cancer, anti-epileptic, anti-osteolytic, and anti-Alzheimer's effects. These findings align with the traditional applications of *C. petasites* in Thai traditional medicine and hold potential for advancing drug development, especially in addressing well-established medical conditions.

REFERENCES

Ambothi K, Prasad NR, Balupillai A. (2015) Ferulic acid inhibits UVB-radiation induced photocarcinogenesis through modulating inflammatory and apoptotic signaling in Swiss albino mice. *Food Chem Toxicol*, 82: 72–78.

Banerjee K, Mandal M. (2015) Oxidative stress triggered by naturally occurring flavone apigenin results in senescence and chemotherapeutic effect in human colorectal cancer cells. *Redox Biol*, 5: 153–162.

Campo G, Pavasini R, Biscaglia S, Ferri A, Andrenacci E, Tebaldi M, Ferrari R. (2015) Platelet aggregation values in patients with cardiovascular risk factors are reduced by verbascoside treatment. A randomised study. *Pharmacol Res,* 97: 1–6.

Chiang LC, Ng LT, Chiang W, Chang MY, Lin CC. (2003) Immunomodulatory activities of flavonoids, monoterpenoids, triterpenoids, iridoid glycosides and phenolic compounds of Plantago species. *Planta Med,* 69: 7, 600–604.

Clavin M, Gorzalczany S, Macho A, Munoz E, Ferraro G, Acevedo C, Martino V. (2007) Anti-inflammatory activity of flavonoids from Eupatorium arnottianum. *J Ethnopharmacol,* 112: 3, 585–589.

Clerodendrum petasites S. Moore [Internet]. The International Plant Names Index (IPNI). 2005 [cited October 2012]. Available from: http://www.ipni.org/ipni/idPlantNameSearch.do;jsessionid=CE5E067923B8D8124E22786A8BA23391?id=862308-1&back_page=%2Fipni%2FeditAdvPlantNameSearch.do%3Bjsessionid%3DCE5E067923B8D8124E22786A8BA23391%3Ffind_infragenus%3D%26find_isAPNIRecord%3Dtrue%26find_geoUnit%3D%26find_includePublicationAuthors%3Dtrue%26find_addedSince%3D%26find_family%3D%26find_genus%3Dclerodendrum%26find_sortByFamily%3Dtrue%26find_isGCIRecord%3Dtrue%26find_infrafamily%3D%26find_rankToReturn%3Dall%26find_publicationTitle%3D%26find_authorAbb.

Gao H, Wang H, Peng J. (2014) Hispidulin induces apoptosis through mitochondrial dysfunction and inhibition of P13k/Akt signalling pathway in HepG2 cancer cells. *Cell Biochem Biophys,* 69: 1, 27–34.

Gao H, Xie J, Peng J, Han Y, Jiang Q, Han M, Wang C. (2015a) Hispidulin inhibits proliferation and enhances chemosensitivity of gallbladder cancer cells by targeting HIF-1alpha. *Exp Cell Res,* 332: 2, 236–246.

Gao L, Peng XM, Huo SX, Liu XM, Yan M. (2015b) Memory enhancement of acteoside (verbascoside) in a senescent mice model induced by a combination of d-gal and AlCl. *Phytother Res,* 29: 8, 1131–1136.

Gil B, Sanz MJ, Terencio MC, Ferrandiz ML, Bustos G, Paya M, Gunasegaran R, Alcaraz MJ. (1994) Effects of flavonoids on Naja naja and human recombinant synovial phospholipases A2 and inflammatory responses in mice. *Life Sci,* 54: 20, 333–338.

Guven M, Aras AB, Akman T, Sen HM, Ozkan A, Salis O, Sehitoglu I, Kalkan Y, Silan C, Deniz M, Cosar M. (2015) Neuroprotective effect of p-coumaric acid in rat model of embolic cerebral ischemia. *Iran J Basic Med Sci,* 18: 4, 356–363.

Hazekamp A, Verpoorte R, Panthong A, Hazekamp A, Verpoorte R, Panthong A. (2001) Isolation of a bronchodilator flavonoid from the Thai medicinal plant Clerodendrum petasites. *J Ethnopharmacol,* 78: 1, 45–49.

He L, Wu Y, Lin L, Wang J, Wu Y, Chen Y, Yi Z, Liu M, Pang X. (2011) Hispidulin, a small flavonoid molecule, suppresses the angiogenesis and growth of human pancreatic cancer by targeting vascular endothelial growth factor receptor 2-mediated PI3K/Akt/mTOR signaling pathway. *Cancer Sci,* 102: 1, 219–225.

Huang WJ, Lee HJ, Chen HL, Fan PC, Ku YL, Chiou LC. (2015) Hispidulin, a constituent of Clerodendrum inerme that remitted motor tics, alleviated methamphetamine-induced hyperlocomotion without motor impairment in mice. *J Ethnopharmacol,* 166: 18–22.

Islas MS, Naso LG, Lezama L, Valcarcel M, Salado C, Roura-Ferrer M, Ferrer EG, Williams PA. (2015) Insights into the mechanisms underlying the antitumor activity of an oxidovanadium (IV) compound with the antioxidant naringenin. Albumin binding studies. *J Inorg Biochem,* 149: 12–24.

Jing W, Chunhua M, Shumin W. (2015) Effects of acteoside on lipopolysaccharide-induced inflammation in acute lung injury via regulation of NF-kappaB pathway in vivo and in vitro. *Toxicol Appl Pharmacol,* 285: 2, 128–135.

Kavvadias D, Sand P, Youdim KA, Qaiser MZ, Rice-Evans C, Baur R, Sigel E, Rausch WD, Riederer P, Schreier P. (2004) The flavone hispidulin, a benzodiazepine receptor ligand with positive allosteric properties, traverses the blood-brain barrier and exhibits anticonvulsive effects. *Br J Pharmacol,* 142: 5, 811–820.

Kim MC, Kim SJ, Kim DS, Jeon YD, Park SJ, Lee HS, Um JY, Hong SH. (2011) Vanillic acid inhibits inflammatory mediators by suppressing NF-kappaB in lipopolysaccharide-stimulated mouse peritoneal macrophages. *Immunopharmacol Immunotoxicol*, 33: 3, 525–532.

Kim SJ, Kim MC, Um JY, Hong SH. (2010) The beneficial effect of vanillic acid on ulcerative colitis. *Molecules*, 15: 10, 7208–7217.

Klaiklay S. (2009) Chemical constituents from the twigs of Garcinia hombroniana, the leaves of Garcinia prainiana and the roots of Clerodendrum petasites S. Moore [MSc. Thesis]. Thailand: Prince of Songkla University.

Kumar B, Gupta SK, Srinivasan BP, Nag TC, Srivastava S, Saxena R, Jha KA. (2013) Hesperetin rescues retinal oxidative stress, neuroinflammation and apoptosis in diabetic rats. *Microvasc Res*, 87: 65–74.

Kwon EY, Jung UJ, Park T, Yun JW, Choi MS. (2015) Luteolin attenuates hepatic steatosis and insulin resistance through the interplay between the liver and adipose tissue in mice with diet-induced obesity. *Diabetes*, 64: 5, 1658–1669.

Lin TY, Lu CW, Wang CC, Lu JF, Wang SJ. (2012) Hispidulin inhibits the release of glutamate in rat cerebrocortical nerve terminals. *Toxicol Appl Pharmacol*, 263: 2, 233–243.

Lin YC, Hung CM, Tsai JC, Lee JC, Chen YL, Wei CW, Kao JY, Way TD. (2010) Hispidulin potently inhibits human glioblastoma multiforme cells through activation of AMP-activated protein kinase (AMPK). *J Agric Food Chem*, 58: 17, 9511–9517.

Nepal M, Choi HJ, Choi BY, Yang MS, Chae JI, Li L, Soh Y. (2013) Hispidulin attenuates bone resorption and osteoclastogenesis via the RANKL-induced NF-kappaB and NFATc1 pathways. *Eur J Pharmacol*, 715: 96–104.

Panthong A, Kanjanapothi D, Taesotikul T, Wongcome T, Reutrakul V. (2003) Anti-inflammatory and antipyretic properties of Clerodendrum petasites S. Moore. J *Ethnopharmacol*, 85: 151–156.

Panthong A, Kanjanapothi D, Taylor W. (1986) Ethnobotanical review of medicinal plants from Thai traditional books, Part I: Plants with anti-inflammatory, anti-asthmatic and antihypertensive properties. *J Ethnopharmacol*, 18: 213–228.

Peng XM, Gao L, Huo SX, Liu XM, Yan M. (2015) The mechanism of memory enhancement of acteoside (verbascoside) in the senescent mouse model induced by a combination of d-gal and AlCl. *Phytother Res*, 29: 1137–1144.

Pesce M, Franceschelli S, Ferrone A, De Lutiis MA, Patruno A, Grilli A, Felaco M, Speranza L. (2015) Verbascoside down-regulates some proinflammatory signal transduction pathways by increasing the activity of tyrosine phosphatase SHP-1 in the U937 cell line. *J Cell Mol Med*, 19: 1548–1556.

Rafacho BP, Stice CP, Liu C, Greenberg AS, Ausman LM, Wang XD. (2015) Inhibition of diethylnitrosamine-initiated alcohol-promoted hepatic inflammation and precancerous lesions by flavonoid luteolin is associated with increased sirtuin 1 activity in mice. *Hepatobiliary Surg Nutr,* 4: 124–134.

Shen Y, Tian P, Li D, Wu Y, Wan C, Yang T, Chen L, Wang T, Wen F. (2015) Chrysin suppresses cigarette smoke-induced airway inflammation in mice. *Int J Clin Exp Med*, 8: 2001–2008.

Shikanga EA, Combrinck S, Regnier T. (2010) South African Lippia herbal infusions: Total phenolic content, antioxidant and antibacterial activities. *S Afr J Bot*, 76: 567–571.

Singh JC, Kakalij RM, Kshirsagar RP, Kumar BH, Komakula SS, Diwan PV. (2015) Cognitive effects of vanillic acid against streptozotocin-induced neurodegeneration in mice. *Pharm Biol,* 53: 630–636.

Singharachai C, Palanuvej C, Kiyohara H, Yamada H, Ruangrungsi N. (2011) Pharmacognostic specification of five root species in Thai traditional medicine remedy: Ben-Cha-Lo-Ka-Wi-Chian. *Phcog J,* 3: 1–11.

Taha MM, Salga MS, Ali HM, Abdulla MA, Abdelwahab SI, Hadi AH. (2012) Gastroprotective activities of Turnera diffusa Willd. ex Schult. revisited: Role of arbutin. *J Ethnopharmacol*, 141: 273–281.

Thai traditional medical textbook: Paet-Ta-Ya-Saat-Song-Kror (2007). conservative ed. The Rehabilitation Foundation for Thai Traditional Medicine and Ayurved Thamrong School, Bangkok, Thailand.

Thitilertdecha P, Guy RH, Rowan MG. (2014a) Characterisation of polyphenolic compounds in Clerodendrum petasites S. Moore and their potential for topical delivery through the skin. *J Ethnopharmacol*, 154: 400–407.

Thongchai W, Liawruangrath B, Liawruangrath S. (2007) High-performance liquid chromatographic determination of arbutin in skin-whitening creams and medicinal plant extracts. *J Cosmet Sci*, 58: 35–44.

Xie J, Gao H, Peng J, Han Y, Chen X, Jiang Q, Wang C. (2015) Hispidulin prevents hypoxia-induced epithelial-mesenchymal transition in human colon carcinoma cells. *Am J Cancer Res*, 5: 1047–1061.

Yang JM, Hung CM, Fu CN, Lee JC, Huang CH, Yang MH, Lin CL, Kao JY, Way TD. (2010) Hispidulin sensitises human ovarian cancer cells to TRAIL-induced apoptosis by AMPK activation leading to Mcl-1 block in translation. *J Agric Food Chem,* 58: 10020–10026.

Yen HR, Liu CJ, Yeh CC. (2015) Naringenin suppresses TPA-induced tumor invasion by suppressing multiple signal transduction pathways in human hepatocellular carcinoma cells. *Chem Biol Interact*, 235: 1–9.

Ying Chang C, Yu Lin T, Wei Lu C, Chuan Wang C, Chou Wang Y, Chou SS, Janc Wang S. (2015) Apigenin, a natural flavonoid, inhibits glutamate release in the rat hippocampus. *Eur J Pharmacol*, 762: 72–81.

Yrbas ML, Morucci F, Alonso R, Gorzalczany S. (2015) Pharmacological mechanism underlying the antinociceptive activity of vanillic acid. *Pharmacol Biochem Behav*, 132: 88–95.

Zhou L, Feng Y, Jin Y, Liu X, Sui H, Chai N, Chen X, Liu N, Ji Q, Wang Y, Li Q. (2014a) Verbascoside promotes apoptosis by regulating HIPK2-p53 signaling in human colorectal cancer. *BMC Cancer*, 14: 747.

6 The Medicinal Properties of Resveratrol, a Polyphenol Component of the Thai Silkworms' Plant *Morus alba*

Alessandra Ferraresi, Chiara Vidoni, and Ciro Isidoro

6.1 INTRODUCTION: RESVERATROL

The 3,4',5-trihydroxystilbene, commonly known as Resveratrol (RV), is a phenolic compound naturally present in a variety of fruits and more abundantly in grape skin (and therefore can be found in wines, especially in red wines) and in berries (which include blueberries, raspberries, strawberries, and mulberry). The red wine may contain up to approximately 14 mg/L of trans-RV.

So far, at least 72 plants and fruits have been reported to contain RV to some extent (Soleas et al., 1997), the actual content being dependent on several factors such as the geographical origin, the climate, the soil composition, and the cultivation and post-cultivation treatments.

It is believed that nature provided the skin of these fruits with RV to prevent them from fungal infection and to preserve the fruit from UV damage. RV is, in fact, a phytoalexin with antimicrobial properties, and it is an antioxidant as well. Thus, avoiding prolonged exposure of the fruit to light is crucial to preserve its medicinal properties, as it could convert the active trans-isomer into the inactive *cis* isomer.

RV was first isolated in 1940 by the Japanese chemist Michio Takaoka from the roots of the flower "white hellebore" (*Veratrum grandiflorum*, commonly known as the Christmas white rose). Subsequently, RV was found in many other plants and fruits, such as peanuts, grapevines, and berries.

Though traditionally, Thai people do not drink wine regularly (compared to European populations), they may introduce a dietary quantity of RV through peanuts and berry juices. A plant endemic to Thailand rich in RV and, mainly, its derivatives

DOI: 10.1201/9781003508724-6

is the *Morus alba* (commonly known as "white" mulberry), which is native to Northern China (Iqbal et al., 2012; Jin et al., 2002). The leaves of this plant are well known worldwide as the preferred feedstock for the silkworm, and this fact explains the widespread cultivation of *Morus alba* throughout the warm temperate world (yet, for the silk industry, a fruitless clone of *M. alba* was developed in the United States of America).

Mulberry leaves (Figure 6.1) were well known to children in Western countries until 50 years ago because they used to raise silkworms in a box to learn how nature makes silk. The practice of raising silkworms on mulberry leaves is still in place in Thailand to produce Thai silk (Figure 6.2).

Another potential source of RV is the fruit of the *Antidesma thwaitesianum*, a Thai berry known in Thailand as Ma-Mao or Mao Luang. This plant belongs to the *angiosperm Phyllanthaceae* and can reach 30 metres tall. It is native to Asia and

FIGURE 6.1 Photograph of mulberry leaves (*Morus alba*). Image kindly provided by Mr James Phungphol.

FIGURE 6.2 The process of manufacturing Thai silk from the cocoons of silkworms.

Southeast Asia, where it is present in a variety of local species in the Philippines (here, the species *A. bunius* is known as Bignay), Vietnam, Taiwan, China, Malaysia, and Thailand (here known as Thai berry), and many other Asian countries.

6.2 THE USE OF RESVERATROL-CONTAINING NATURAL PRODUCTS IN TRADITIONAL MEDICINE

RV gained its worldwide reputation as a dietary product with medicinal properties only recently. Knowledge of the presence of RV in grapevines dates to 1976. Subsequently, in 1992, it was also reported in wine. In the 1990s, it was found that red wine (moderate) drinkers in the South of France and North-West of Italy (and other Mediterranean countries) were statistically less disposed to develop atherosclerosis and heart attack, despite the presence of risk factors such as little exercise and a high-fat diet (Biagi et al., 2015; Catalgol et al., 2012). As this observation became known, the "French paradox" was associated with low blood pressure. Soon, it became clear that it was not due to the alcohol present in the wine but instead to a substance capable of inducing vascular dilation: this substance was found to be *trans*-RV. The relative content of RV in red wine (up to approximately 12 mg/L) varies with the type of grape (or mixture of grapes) and the geographical region of cultivation (Goldberg et al., 1995). However, the curative properties of natural extracts containing this substance were already known in ancient Asian medicine.

In 1963, the group of Makimoto isolated RV along with polydatin (its beta-glucoside) from *Polygonum cuspidatum* (Nonomura et al., 1963). These two products were found to be the most active components of the dried roots of this plant, also known as Fallopia japonica (or Japanese knotweed because of its resemblance to bamboo). Curiously, while in Japan, this plant is held in high regard, it is hated in the rest of the world, and its growth is hindered because it causes damage to the land and gardens.

In Japanese, the plant is called Itadori, while in Chinese medicine, it is called Huzhang. The Chinese meaning of Huzhang is "tiger staff," while the Japanese meaning for Itadori is "remove the pain." The popular name justifies the interest of scientists. The Japanese chemists were interested in finding the bioactive components of this plant because the dried powder of its root was renowned in traditional Japanese and Chinese medicine as a potent remedy for several pathological afflictions.

Ko-jo-kon, the folk name of the powder, was used in ancient times to treat allergic and inflammatory reactions, fungal infections, bacterial infections (e.g., gonorrhoea), and obesity and diabetes (what we would now call "metabolic syndromes").

In traditional Chinese medicine, the fruit and the bark of *Polygonum cuspidatum are used*, but not the roots. As mentioned above, RV is also present in white mulberry (*Morus alba*), which is imported into Thailand to produce Thai silk. It is to be noted, however, that the relative content of RV and oxy-RV differs in the different portions of the plant (leaves, bark, and root bark), and it is greatly influenced by the soil, the cultivation, and the collecting season (Zhou et al., 2013).

*Morus alb*a is well-known in Chinese traditional medicine. The root bark of this mulberry is the basis for the Sang-Bai-Pi herbal medicine (Hano et al., 1984). Sang-Bai-Pi is used in traditional Chinese medicine (6–15 g/day) to treat lung and throat inflammatory pathologies and relieve cough, asthma, and dyspnoea. This herbal medicine is so effective that it is commercialised worldwide by many Chinese companies in granules for tea. The tea is naturally sweet and is effective in promoting urination to fight fever, headache, and inflammatory conditions.

The fruit was given to treat diabetes (despite its sweetness) and constipation, and it was recommended as an anti-ageing remedy to prevent the premature greying of the hairs. It is worth noting that RV has confirmed that all these properties will be elicited. As mentioned earlier, RV is also present in the Thai berry *Antidesma thwaitesianum,* known locally as Ma-Mao or Mao Luang.

The antioxidant and antimicrobial activities of the Ma-Mao fruit extracts (Figure 6.3) have been intensely investigated, and they were found to differ concerning the content of polyphenols, which varied depending on the method used for the preparation (Dechayont et al., 2012). The extract from the fruit contains phenols and flavonoids, and it was shown to elicit cytotoxic effects on cancer cells *in vitro* (Pinturson et al., 2015) and anti-inflammatory effects (Puangpronpitag et al., 2008, 2011). Yet, traditional Thai medicine does not use the pure extract from this fruit, which is used in Thailand to make juice, jams, and wine (Figure 6.3).

It is worth noting that the Ma-Mao juice is advertised and sold as a rejuvenating elixir. It is indeed full of anthocyanins with antioxidant properties, and it also contains RV, which has been proven to prolong the lifespan of several experimental organisms (see below).

In Thai traditional medicine, other herbal remedies are used as rejuvenators. Among these, the most popular that have been shown to possess antioxidant, anti-inflammatory, and anticancer properties are Jatu-Pha-La-Ti-Ga, Tri-Sa-Mo, and Tri-Su-Ra-Phon (Chusri et al., 2015). It has not been determined if these extracts

FIGURE 6.3 Photograph of a bottle of Mao juice.

contain RV, though it is very likely considering their phenol content and anti-ageing properties.

The plant sources, the technologies for extraction and purification, the stability, and food and medical applications of RV have recently been reviewed (Tian and Liu, 2020).

6.3 ABSORPTION AND METABOLISM OF RESVERATROL

The hydroxyl groups (three in RV and four in Piceatannol or oxy-RV) confer the molecule amphipathic characteristics that partially facilitate its solubility in body fluids. Yet, its cellular absorption is facilitated in the presence of a small quantity of alcohol (Figure 6.4) (Neves et al., 2012).

On the contrary, the methoxy-derivative Pterostilbene is much more efficiently absorbed *in vivo*. Once absorbed in humans, Pterostilbene is metabolised into a sulphate-derivative, while RV is metabolised into a glucuronide conjugate. Pterostilbene is considered, for this reason, more potent than RV, though the studies do not fully support this contention. The different absorption rates and the diverse metabolism and distribution in the body may explain why the effects elicited by these two substances do not overlap completely. RV's bioavailability, absorption, and metabolism have been extensively investigated in various animal models and humans (Gambini et al., 2015).

6.4 BIOCHEMICAL PROPERTIES AND ACTIVITIES OF RESVERATROL AND ITS DERIVATIVES/ANALOGOUS

The molecular structure of RV and its derivatives can easily explain most of this compound's chemical and biological properties. As a polyphenol, it possesses antioxidant properties, and as an oestrogen mimetic, it has the property of a phytoestrogen.

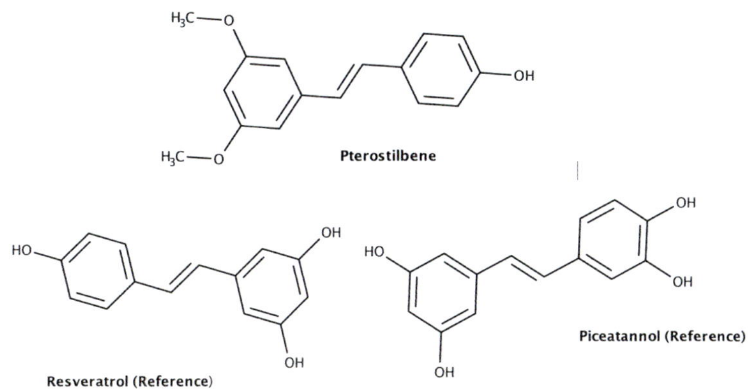

FIGURE 6.4 Structures of Resveratrol and its derivates.

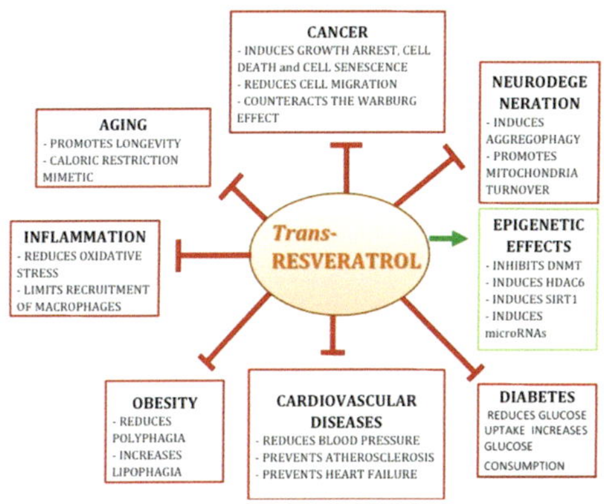

FIGURE 6.5 Schematic representation of pleiotropic effects of Resveratrol.

Besides, RV can interact directly with specific proteins and interfere with their activity and function in the cell. In addition, RV can modulate the transcription of genes and, epigenetically, can control protein expression. These chemical characteristics confer RV pleiotropic functions in biological systems (Figure 6.5) (Calamini et al., 2010).

6.4.1 Antioxidant Activity

The chemical polyphenolic structure of RV accounts for its ability to scavenge free radicals (Gülçin et al., 2010). Besides, RV actively prevents the generation of reactive oxygen species by inhibiting pro-oxidant enzymes.

For instance, RV can bind and inhibit the cyclooxygenases COX-1 and COX-2 and the hydroperoxides (Jang et al., 1997). Yet, the enzyme that RV tightly binds with the highest affinity is the FAD-dependent quinone oxidoreductase QR2 (NQO2) (St John et al., 2013). QR2/NQO2 activity leads to the generation of anion superoxide, which can harm the cell as it may cause DNA mutations, potentially leading to cancer (Slane et al., 2006). In addition, very recently, it has been shown that oxidative stress generated by QR2/NQO2 can inhibit protective autophagy in astroglial and neuronal cells (Janda et al., 2015). However, it is to be noted that the sulphate and glucuronyl metabolites of RV that are produced in the animal bind less efficiently to QR2.

In addition, RV was shown to decrease intracellular ROS levels through the downregulation of NADPH oxidase subunits NOX1 and NOX4 (He et al., 2015). Finally, RV shuts down oxidative stress by activating the transcription factor Nrf-2, which leads to the synthesis of glutathione peroxidase-2 (Soeur et al., 2015).

6.4.2 Inhibition of the PI3K-AKT-mTOR Pathway

Tyrosine kinase receptors activate the class IA phosphoinositide 3 kinase. It is involved in mitogenic signalling and metabolic control by activating the AKT-mTOR downstream pathway that controls cell survival, growth, and protein synthesis. RV has been shown to shut down this pathway by directly inhibiting the class IA PI3K catalytic subunits p110α and p110β (Fröjdö et al., 2007). This activity has many implications in several cellular functions, such as in the control of cancer cell proliferation (Sui et al., 2014) and glucose uptake (Chi et al., 2007).

6.4.3 Inhibition of Melanogenesis

Oxy-RV inhibits the activity of tyrosinase and, therefore, can limit the biosynthesis of melanin (Kim et al., 2002; Liu et al., 2015). This property prompts its use as a cosmetic drug to reduce sun-induced darkening of the skin and as a medical drug for treating skin hyperpigmentation disorders.

6.4.4 Inhibition of Polymerase Activity

Stivala et al. (2001) first showed that the antiproliferative activity of Pterostilbene (the methoxy derivative of RV) correlated with its ability to inhibit *in vitro* the DNA polymerases alpha and delta (Stivala et al., 2001). Later, this same group investigated the molecular mechanisms of this inhibition in detail and determined the specific structural determinants of the RV molecule responsible for such selective inhibition of different mammalian polymerases (Locatelli et al., 2005).

6.4.5 Activation of the Nitric Oxide Synthase

Since the description of the molecular basis of the French Paradox, it was reported that RV could induce a relaxation of the blood vessel wall through the induction of

NO. Several mechanisms are involved in this effect, including the upregulation of eNOS expression in the endothelial cells and the stimulation of its enzymatic activity (Xia et al., 2014).

6.4.6 Resveratrol as a Phytoestrogen

Given its structural resemblance to natural oestrogens (see the similarity with 17-β2-estrogen or oestradiol), RV can bind the oestrogen receptors (ER) (Gehm et al., 1997). However, while the binding of E2 to its ER will lead to the transcription of target genes, the binding of RV to ER is sterile regarding gene expression. Therefore, RV can be used as a competitor for natural oestrogens to halt their activity. Indeed, the hope is that RV, like other natural phytoestrogens, can effectively treat oestrogen-responsive cancers (Hwang and Choi, 2015).

6.4.7 Resveratrol as an Inducer of Autophagy

Autophagy (literally "self-eating") refers to the lysosomal degradation of cellular constituents and organelles. The cell may use this process for the physiological turnover of its components (basal autophagy) or to eliminate damaged, altered, or redundant cellular components (induced autophagy). There are three types of autophagy: macro-autophagy, micro-autophagy, and chaperone-mediated autophagy (Feng et al., 2015). The autophagy process has a significant role in cell homeostasis by controlling the cellular mass. It is, therefore, interconnected with other pathways involved in cell life, including protein synthesis and cell survival or cell death (Anding and Baehrecke, 2015). The pivotal role of autophagy in practically all aspects of cell metabolism explains its importance in organ function and organism homeostasis and, consequently, in so many pathophysiological processes (Ravikumar et al., 2010; Klionsky et al., 2021).

RV can regulate autophagy through the AMPK pathway (Ha et al., 2015). The first evidence showing the induction of autophagy by RV was made in ovarian cancer cells by Opipari and colleagues (Opipari et al., 2004). Subsequently, it was demonstrated that RV could induce autophagy-dependent cell death in colorectal cancer cells (Trincheri et al., 2008). As a caloric- and protein-restriction mimetic, RV has been shown to induce autophagy more strongly than amino acid and serum starvation (Ferraresi et al., 2017a). It is now well established that most of the beneficial effects elicited by RV, including the protection of cardiovascular and neurological systems, anticancer and anti-ageing effects, and the curative effects on metabolic diseases, are mediated through autophagy (Mariño et al., 2014; Hasima et al., 2014; Albani et al., 2010; Morselli et al., 2010; Gurusamy et al., 2010; Klionsky et al., 2021).

6.4.8 Resveratrol as an Epigenetic Regulator

Epigenetics refers to those hereditary changes in gene expression that do not involve any alteration in the DNA sequence controlling the transcription or the sequence coding for the protein. Essentially, the expression of the gene is epigenetically regulated by three mechanisms that involve the modulation of the level of methylation/demethylation of the promoter (which regulates the accessibility to the transcription factors), the level of acetylation/deacetylation of the histones (which regulates the opening of the chromatin structure and the degree of DNA condensation), and the synthesis of long- and micro- non-coding RNAs that control the transcription and the translation of the mRNAs.

Thus, gene expression can be epigenetically modulated by interfering with the activity of DNA-methyl transferases (DNMT), histone deacetylases (HDAC), and the synthesis of LncRNAs and microRNAs.

RV has been shown to possess these activities. HDAC-mediated histone deacetylation determines an increase in the affinity of histones to DNA, which then becomes more condensed and unavailable for transcription. In contrast, histone acetylation by histone-acetylases (HAC) elicits the opposite effect, allowing gene transcription. RV can modulate these activities (Gao and Tollefsbol, 2015; Vahid et al., 2015; Stefanska et al., 2012). RV can also increase the accessibility of the promoter by inhibiting DNMTs (Singh et al., 2013).

Finally, RV can epigenetically influence epigenetically the level of gene expression in protein by modulating the synthesis of long non-coding RNAs and microRNAs, which in turn can modulate the transcription and translation of mRNAs (Kumar et al., 2015; Latruffe et al., 2015; Phuah and Nagoor, 2014; Ferraresi et al., 2017b; Vallino et al., 2020; Asemi et al., 2023).

6.5 PATHOPHYSIOLOGICAL ACTIVITIES OF RESVERATROL

6.5.1 Anti-Inflammatory

RV can be proposed for the cure of transitory inflammatory conditions as well as chronic inflammatory diseases (e.g., rheumatic diseases) based on its ability to inhibit the expression of cyclooxygenase 1 (COX1) and cyclooxygenase 2 (COX2) (Kundu et al., 2006) and of prostaglandin E2 (PGE2) and ROS (Candelario-Jalil et al., 2007; Kim et al., 2007).

6.5.2 Anti-Ageing

RV can trigger the expression of Sirtuins (SirTs), genes associated with longevity (Miyazaki et al., 2008; Borra et al., 2005; Kaeberlein et al., 2005; Jiang et al., 2008; Guarente et al., 2007). RV has been shown to extend the lifespan in several organisms, including Drosophila (Griswold et al., 2008), C. elegans (Gruber et al., 2007), and the short-lived fish Northobranchius (Terzibasi et al., 2007). RV was also shown to extend the lifespan of mammals (mice) (Baur et al., 2006). This study showed that a high-calorie diet (60% of calories from fat)–induced obesity, diabetes, and atherosclerosis, eventually reducing the lifespan of middle-aged (1-year-old) mice. In contrast, administering RV (22.4 mg/kg/day) along with the same diet extended the animal's lifespan. These effects were mediated by SIRT1, a gene induced by caloric restriction. By inducing SIRT1 expression, RV elicits the same beneficial effects as a hypo-caloric diet, and therefore, it is often referred to as a "restriction caloric mimetic" (Nisoli et al., 2005). A comprehensive and detailed description of the anti-ageing mechanisms triggered by RV has been recently reviewed (Pyo et al., 2020).

6.6 RESVERATROL FOR TREATING HUMAN DISEASES

6.6.1 Cardiovascular Diseases

Epidemiological, *in vitro*, and *in vivo* preclinical studies have provided convincing evidence that RV elicits beneficial effects in cardiovascular-related pathologies, including atherosclerosis, hypertension, ischaemic heart disease, and heart failure (Khurana et al., 2013).

Several studies have shown that low to moderate intake (200 g/day) of RV-containing wine decreases the risk of death from cardiovascular diseases (Gronbaek et al., 1995; Klatsky et al., 2003).

The beneficial effects of RV on the cardiovascular system are mediated through the activation of the deacetylase SIRT1 (silent information regulator 1), the induction of AMPK-dependent autophagy, NO-mediated vasodilation, and the general inhibition of ROS production.

6.6.2 Neurodegenerative Diseases

Neurodegenerative diseases arise from neuronal cell loss in brain areas caused by exaggerated oxidative stress and the accumulation of undigested anomalous proteins. RV can dampen oxidative stress and, by inducing autophagy, can help the neurons get rid of the protein aggregates. Therefore, it is conceivable that RV could and should have neuroprotective properties. Indeed, RV has shown promising therapeutic potential in *in vitro* cell models and *in vivo* animal models of several human neurodegenerative disorders (Tellone et al., 2015).

RV was shown to be able to prevent the formation of toxic Aβ oligomers and protofibrillar intermediates, delaying Aβ toxicity in cultured and animal models of Alzheimer's disease (Savaskan et al., 2003; Feng et al., 2009; Rege et al., 2015). On these bases, RV has been proposed for the therapy of Alzheimer's disease (Braidy et al., 2016), and the results of a first clinical trial have been recently published (Turner et al., 2015).

RV exerted a neuroprotective effect in various *in vitro* and animal models of Huntington's Disease by reducing the oxidative stress and toxicity of the mutant Huntingtin protein through the induction of autophagy (Vidoni et al., 2018), of the ERK pathway (Maher et al., 2011) and of SIRT1 (Ho et al., 2010). Similarly, RV was shown to protect dopaminergic neurons from the toxicity induced by parkinsonian neurotoxins, such as rotenone, 1-methyl-4-phenyl-1,2,3,6-tetrahydropyridine, and 6-hydroxy-dopamine, both in *in vitro* (Chao et al., 2008; Ferretta et al., 2014; Lin et al., 2015) and in animal models (Anandhan et al., 2010; Jin et al., 2008). Finally, a recent report showed the beneficial effects of RV administration in the SOD1G93A mouse model of amyotrophic lateral sclerosis (Song et al., 2014).

6.6.3 Cancer

The potential chemopreventive and therapeutic properties of RV in cancer have been recently reviewed by Seyed et al. (Seyed et al., 2016; Ren et al., 2021).

RV has been shown to inhibit all steps of cancer development in various cell and animal models (Jang and Pezzuto, 1999). A seminal work by the group of Pezzuto showed that RV could act as an anti-mutagen agent by inducing phase II drug-metabolizing enzymes, as an anti-promotion agent by inhibiting cyclooxygenase and hydroperoxidase functions, and as a pro-differentiating agent of cancer cells (Jang et al., 1997). RV could also prevent the progression of preneoplastic lesions (Jang et al., 1997) and the development of tumours in xenograft models (van Ginkel et al., 2007) and chemical carcinogenesis models (Tessitore et al., 2000). *In vitro,* RV has been shown to inhibit proliferation and cell migration, to induce differentiation and cell death in a variety of cancer cell types, including human colorectal cancer cells (Trincheri et al., 2007), breast cancer cells (Venkatadri et al., 2016), prostate cancer (Selvaraj et al., 2015), human glioblastoma cells (Castino et al., 2011; Xiong et al., 2016), ovarian cancer cells (Ferraresi, Esposito et al. 2021; Ferraresi et al., 2017b), cholangiocarcinoma cells (Thnogchot et al., 2018), and many others. RV has been shown to induce cancer cell dormancy, thus preventing cytokine-induced relapse (Esposito, Ferraresi et al., 2022). In preclinical animal models, resveratrol has been shown to inhibit the *in vivo* growth of lung cancer (Savio et al., 2022), ovarian cancer (Tan et al., 2016; Zhong et al., 2019), and cholangiocarcinoma (Thongchot et al., 2024), among others.

6.7 CLINICAL TRIALS WITH RESVERATROL

A first-phase I pharmacokinetic study of escalation of RV at a single dose of up to 5 g was performed in 2007, and no adverse effects were reported (Boocock et al., 2007). Subsequent studies showed that up to 1 g/day for three months does not elicit undesired effects (Gescher et al., 2013), even in obese, older subjects (Anton et al., 2014).

A more recent study confirmed that RV at doses of ≥0.5 g/day for long periods is well tolerated, with only moderate and reversible abdominal discomfort (Cottart et al., 2014)

There are 217 clinical trials listed on the NIH website for clinical trials (https://clinicaltrials.gov), of which 145 have been concluded. These are currently listed on the website:

https://clinicaltrials.gov/ct2/results?term=RESVERATROL&Search=Search

These studies involve cardiovascular, neurodegenerative disorders (e.g., Alzheimer's disease), metabolic diseases (including diabetes and obesity), Friedrich's ataxia, cancer, memory impairment, chronic renal insufficiency, hypertension, seasonal allergic rhinitis, age-related macular degeneration, osteoporosis, chronic inflammation, dyslipidaemia, mitochondrial function/anti-ageing, endometriosis, and endothelial dysfunction. An in-depth analysis of the clinical studies supporting the beneficial effects of RV in human diseases can be found in recent review articles (Berman et al., 2017; Singh et al., 2019; Zhang et al., 2021; Brown et al., 2024). One important point that deserves particular attention is the choice of the most appropriate dose of RV to administer to obtain the desired effect. RV is, in fact, a hermetic drug that elicits pro-survival effects at low concentrations and pro-death effects at high concentrations (Mukherjee et al., 2010). While *in vitro,* it can be relatively easy to set the range for "Low" and "High" concentrations depending on the cellular type, this is challenging *in vivo* where RV is administered systemically, and pharmacokinetic and pharmacodynamic processes affect the actual amount of RV available for the target organ. The challenge is to define the effective dose to administer to the patient, which depends on the disease and the desired effect (anti-ageing, anti-inflammatory, cytotoxic, pro-survival, etc.) and the characteristics and habits of the patient (diet, microbiome, etc.). For instance, for the adjuvant treatment of cancer patients, the effective dose can be calculated by translating the dose from animal studies (Ferraresi et al., 2024).

REFERENCES

Albani D, Polito L, Signorini A, et al., Neuroprotective properties of resveratrol in different neurodegenerative disorders. 2010. *Biofactors*, 36(5): 370–6.

Anandhan A, Tamilselvam K, Vijayraja D, et al., resveratrol attenuates oxidative stress and improves behaviour in 1 -methyl-4-phenyl-1,2,3,6-tetrahydropyridine (MPTP) challenged mice. 2010. *AnnNeurosci*, 17(3): 113–9.

Anding AL, Baehrecke EH. Autophagy in cell life and cell death. 2015. *Curr Top Dev Biol*, 114: 67–91.

Anton SD, Embry C, Marsiske M, et al., Safety and metabolic outcomes of resveratrol supplementation in older adults: Results of a twelve-week, placebo-controlled pilot study. 2014. *Exp Gerontol*, 57: 181–7.

Asemi R, Rajabpoor Nikoo N, et al., Modulation of long non-coding RNAs by resveratrol as a potential therapeutic approach in cancer: A comprehensive review. 2023. *Pathol Res Pract,* 246: 154507. doi: 10.1016/j.prp.2023.154507. Epub 2023 May 6. PMID: 37196467.

Baur JA, Pearson KJ, Price NL, et al., Resveratrol improves health and survival of mice on a high-calorie diet. 2006. *Nature*, 444(7117): 337–42.

Berman AY, Motechin RA, Wiesenfeld MY, et al., The therapeutic potential of resveratrol: A review of clinical trials. 2017. *NPJ Precis Oncol*, 1: 35. doi: 10.1038/s41698-017-0038-6.

Biagi M, Bertelli AA. Wine, alcohol and pills: What future for the French paradox? 2015. *Life Sci*, 15(131): 19–22.

Boocock DJ, Faust GE, Patel KR, et al., Phase I dose escalation pharmacokinetic study in healthy volunteers of resveratrol, a potential cancer chemopreventive agent. 2007. *Cancer Epidemiol Biomarkers Prev*, 16(6): 1246–52.

Borra MT, Smith BC, Denu JM. Mechanism of human SIRT1 activation by resveratrol. 2005. *J Biol Chem*, 280(17): 17187–95.

Braidy N, Jugder BE, Poljak A, et al., Resveratrol as a potential therapeutic candidate for the treatment and management of Alzheimer's disease. 2016. *Curr Top Med Chem*, 16(17): 1951–60.

Brown K, Theofanous D, Britton RG, et al., Resveratrol for the management of human health: How far have we come? A systematic review of resveratrol clinical trials to highlight gaps and opportunities. 2024. *Int J Mol Sci,* 25(2): 747. doi: 10.3390/ijms25020747.

Calamini B, Ratia K, Malkowski MG, et al., Pleiotropic mechanisms facilitated by resveratrol and its metabolites. 2010. *Biochem J*, 429(2): 273–82.

Candelario-Jalil E, de Oliveira AC, Gräf S, et al., Resveratrol potently reduces prostaglandin E2 production and free radical formation in lipopolysaccharide-activated primary rat microglia. 2007. *J Neuroinflammation*, 4: 1–12.

Castino R, Pucer A, Veneroni R, et al., Resveratrol reduces the invasive growth and promotes the acquisition of a long-lasting differentiated phenotype in human glioblastoma cells. 2011. *J Agric Food Chem*, 59(8): 4264–72.

Catalgol B, Batirel S, Taga Y, et al., Resveratrol: French paradox revisited. 2012. *Front Pharmacol*, 3: 141.

Chao J, Yu MS, Ho YS, et al., Dietary oxyresveratrol prevents parkinsonian mimetic 6-hydroxydopamine neurotoxicity. 2008. *Free Radic Biol Med*, 45(7): 1019–26.

Chi TC, Chen WP, Chi TL, et al., Phosphatidylinositol-3-kinase is involved in the antihyperglycemic effect induced by resveratrol in streptozotocin-induced diabetic rats. 2007. *Life Sci*, 80(18): 1713–20.

Chusri S, Singthong P, Kaewmanee T. Antioxidant, anticancer, and cytotoxic effects of Thai traditional herbal preparations consumed as rejuvenators. 2015. *CyTA - Journal of Food*, 13(1): 40–48.

Cottart CH, Nivet-Antoine V, Beaudeux JL. Review of recent data on the metabolism, biological effects, and toxicity of resveratrol in humans. 2014. *Mol Nutr Food Res,* 58(1): 7–21. doi: 10.1002/mnfr.201200589. Epub 2013 Jun 6. Review.

Dechayont B, Hansakul P, Itharat A. Comparison of antimicrobial, antioxidant activities and total phenolic content of Antidesma thwaitesianum fruit extracts by different methods. 2012. *J Med Assoc Thai,* 95 Suppl 1: S147–53.

Esposito A, Ferraresi A, Salwa A, et al., Resveratrol contrasts IL-6 pro-growth effects and promotes autophagy-mediated cancer cell dormancy in 3D ovarian cancer: Role of miR-1305 and of its target ARH-I. 2022. *Cancers (Basel),* 14(9): 2142. doi: 10.3390/cancers14092142.

Feng Y, Wang XP, Yang SG, et al., Resveratrol inhibits beta-amyloid oligomeric cytotoxicity but does not prevent oligomer formation. 2009. *Neurotoxicology,* 30(6): 986–95.

Feng Y, Yao Z, Klionsky DJ. How to control self-digestion: Transcriptional, post-transcriptional, and post-translational regulation of autophagy. 2015. *Trends Cell Biol*, 25(6): 354–63.

Ferraresi A, Titone R, Follo C, et al., The protein restriction mimetic resveratrol is an autophagy inducer stronger than amino acid starvation in ovarian cancer cells. 2017a. *Mol Carcinog,* 56(12): 2681–91. doi: 10.1002/mc.22711. Epub 2017 Sep 7. PMID: 28856729.

Ferraresi A, Phadngam S, Morani F, et al., Resveratrol inhibits IL-6-induced ovarian cancer cell migration through epigenetic up-regulation of autophagy. 2017b. *Mol Carcinog,* 56(3): 1164–81. doi: 10.1002/mc.22582. Epub 2016 Nov 3. PMID: 27787915.

Ferraresi A, Esposito A, Girone C, et al., Resveratrol contrasts LPA-induced ovarian cancer cell migration and platinum resistance by rescuing Hedgehog-Mediated autophagy. 2021. *Cells,* 10(11): 3213. doi: 10.3390/cells10113213. PMID: 34831435; PMCID: PMC8625920.

Ferraresi A, Thongchot S, Isidoro C. Resveratrol promotes self-digestion to put cancer to sleep. 2024. *J Cancer Prev,* 29(1): 1–5. doi: 10.15430/JCP.24.001.

Ferretta A, Gaballo A, Tanzarella P, et al., Effect of resveratrol on mitochondrial function: Implications in parkin-associated familiar Parkinson's disease. 2014. *Biochim Biophys Acta*, 1842(7): 902–15.

Fröjdö S, Cozzone D, Vidal H, et al., Resveratrol is a class IA phosphoinositide 3-kinase inhibitor. 2007. *Biochem J*, 406(3): 511–8.

Gambini J, Inglés M, Olaso G, et al., Properties of resveratrol: In vitro and in vivo studies about metabolism, bioavailability, and biological effects in animal models and humans. 2015. *Oxid Med Cell Longev*, 2015: 837042.

Gao Y, Tollefsbol TO. Impact of epigenetic dietary components on cancer through histone modifications. 2015. *Curr Med Chem*, 22(17): 2051–64.

Gehm BD, McAndrews JM, Chien PY, et al., Resveratrol, a polyphenolic compound found in grapes and wine, is an agonist for the estrogen receptor. 1997. *Proc Natl Acad Sci U S A*, 94(25): 14138–43.

Gescher A, Steward WP, Brown K. Resveratrol in the management of human cancer: How strong is the clinical evidence? 2013. *Ann N Y Acad Sci*, 1290: 12–20.

Goldberg DM, Yan J, Ng E, et al., A global survey of trans-resveratrol concentrations in commercial wines. 1995. *Am J Enol Vitic*, 46: 159–65.

Griswold AJ, Chang KT, Runko AP, et al., Sir2 mediates apoptosis through JNK-dependent pathways in Drosophila. 2008. *Proc Natl Acad Sci U S A*, 105(25): 8673–8.

Gruber J, Tang SY, Halliwell B. Evidence for a trade-off between survival and fitness caused by resveratrol treatment of Caenorhabditis elegans. 2007. *Ann N Y Acad Sci*, 1100: 530–42.

Guarente L. Sirtuins in aging and disease. 2007. *Cold Spring Harb Symp Quant Biol*, 72: 483–8.

Gülçin I. Antioxidant properties of resveratrol: A structure–activity insight. 2010. *Innov Food Sci Emerg Technol*, 11(1): 210–18.

Gurusamy N, Lekli I, Mukherjee S, et al., Cardioprotection by resveratrol: A novel mechanism via autophagy involving the mTORC2 pathway. 2010. *Cardiovasc Res*, 86(1): 103–12.

Ha J, Guan KL, Kim J. AMPK and autophagy in glucose/glycogen metabolism. 2015. *Mol Aspects Med*, 46: 46–62.

Hano Y, Itoh M, Koyoma N, et al., Constituents of the Chinese crude drug Sang-Bai-Pi (Morus Root Bark) V. structures of three new flavanones, Sanggenons L, M, and N. 1984. *Heterocycles (Sendai)*, 22(8): 1791–800.

Hasima N, Ozpolat B. Regulation of autophagy by polyphenolic compounds as a potential therapeutic strategy for cancer. 2014. *Cell Death Dis*, 5: e1509.

He T, Guan X, Wang S, et al., Resveratrol prevents high glucose-induced epithelial-mesenchymal transition in renal tubular epithelial cells by inhibiting NADPH oxidase/ROS/ERK pathway. 2015. *Mol Cell Endocrinol*, 402: 13–20.

Ho DJ, Calingasan NY, Wille E, et al., Resveratrol protects against peripheral deficits in a mouse model of Huntington's disease. 2010. *Exp Neurol*, 225(1): 74–84.

Hwang KA, Choi KC. Anticarcinogenic effects of dietary phytoestrogens and their chemopreventive mechanisms. 2015. *Nutr Cancer*, 67(5): 796–803.

Iqbal S, Younas U, Sirajuddin, et al., Proximate composition and antioxidant potential of leaves from three varieties of Mulberry (Morus sp.): A comparative study. 2012. *Int J Mol Sci*, 13(6): 6651–64.

Janda E, Lascala A, Carresi C, et al., Parkinsonian toxin-induced oxidative stress inhibits basal autophagy in astrocytes via NQO2/quinone oxidoreductase 2: Implications for neuroprotection. 2015. *Autophagy*, 11(7): 1063–80.

Jang M, Cai L, Udeani GO, et al., Cancer chemopreventive activity of resveratrol, a natural product derived from grapes. 1997. *Science*, 275(5297): 218–20.

Jang M, Pezzuto JM. Cancer chemopreventive activity of resveratrol. 1999. *Drugs Exp Clin Res*, 25(2–3): 65–77.

Jiang WJ. Sirtuins: Novel targets for metabolic disease in drug development. 2008. *Biochem Biophys Res Commun*, 373(3): 341–4.

Jin F, Wu Q, Lu YF, et al., Neuroprotective effect of resveratrol on 6-OHDA-induced Parkinson's disease in rats. 2008. *Eur J Pharmacol,* 600(1–3): 78–82.

Jin W, Na M, An R, et al., Antioxidant compounds from twig of Morus alba. 2002. *Nat Prod Sci,* 8(4): 129–32.

Kaeberlein M, McDonagh T, Heltweg B, et al., Substrate-specific activation of sirtuins by resveratrol. 2005. *J Biol Chem,* 280(17): 17038–45.

Khurana S, Venkataraman K, Hollingsworth A, et al., Polyphenols: Benefits to the cardiovascular system in health and in aging. 2013. *Nutrients,* 5(10): 3779–827.

Kim YA, Kim GY, Park KY, et al., Resveratrol inhibits nitric oxide and prostaglandin E2 production by lipopolysaccharide-activated C6 microglia. 2007. *J Med Food,* 10(2): 218–24.

Kim YM, Yun J, Lee CK, et al., Oxyresveratrol and hydroxystilbene compounds. Inhibitory effect on tyrosinase and mechanism of action. 2002. *J Biol Chem,* 277(18): 16340–4.

Klatsky AL, Friedman GD, Armstrong MA, et al., Wine, liquor, beer, and mortality. 2003. *Am J Epidemiol,* 158(6): 585–95.

Klionsky DJ, Petroni G, Amaravadi RK, et al., Autophagy in major human diseases. 2021. *EMBO J.* 40(19): e108863. doi: 10.15252/embj.2021108863. Epub 2021 Aug 30. PMID: 34459017; PMCID: PMC8488577.

Kumar A, Dhar S, Rimando AM, et al., Epigenetic potential of resveratrol and analogs in preclinical models of prostate cancer. 2015. *Ann N Y Acad Sci,* 1348(1): 1–9.

Kundu JK, Shin YK, Kim SH, et al., Resveratrol inhibits phorbol ester-induced expression of COX-2 and activation of NF-kappaB in mouse skin by blocking IkappaB kinase activity. 2006. *Carcinogenesis,* 27(7): 1465–74.

Latruffe N, Lançon A, Frazzi R, et al., Exploring new ways of regulation by resveratrol involving miRNAs, with emphasis on inflammation. 2015. *Ann N Y Acad Sci,* 1348(1): 97–106.

Lin TK, Chen SD, Chuang YC, et al., Resveratrol partially prevents rotenone-induced neurotoxicity in dopaminergic SH-SY5Y cells through induction of heme oxygenase-1 dependent autophagy. 2014. *Int J Mol Sci,* 15(1): 1625–46.

Liu Q, Kim C, Jo YH, et al., Synthesis and biological evaluation of resveratrol derivatives as melanogenesis inhibitors. 2015. *Molecules,* 20(9): 16933–45.

Locatelli GA, Savio M, Forti L, et al., Inhibition of mammalian DNA polymerases by resveratrol:mechanism and structural determinants. 2005. *Biochem J,* 389(Pt2): 259–68.

Maher P, Dargusch R, Bodai L, et al., ERK activation by the polyphenols fisetin and resveratrol provides neuroprotection in multiple models of Huntington's disease. 2011. *Hum Mol Genet,* 20(2): 261–70.

Mariño G, Pietrocola F, Madeo F, et al., Caloric restriction mimetics: Natural/physiological pharmacological autophagy inducers. 2014. *Autophagy,* 10(11): 1879–82.

Miyazaki R, Ichiki T, Hashimoto T, et al., SIRT1, a longevity gene, downregulates angiotensin II type 1 receptor expression in vascular smooth muscle cells. 2008. *Arterioscler Thromb Vasc Biol,* 28(7): 1263–9.

Morselli E, Maiuri MC, Markaki M, et al., The life span-prolonging effect of sirtuin-1 is mediated by autophagy. 2010. *Autophagy,* 6(1): 186–8.

Mukherjee S, Dudley JI, Das DK. Dose-dependency of resveratrol in providing health benefits. 2010. *Dose Response,* 8(4): 478–500.

Neves AR, Lucio M, Lima JL, et al., Resveratrol in medicinal chemistry: A critical review of its pharmacokinetics, drug-delivery, and membrane interactions. 2012. *Curr Med Chem,* 19(11): 1663–81.

Nisoli E, Tonello C, Cardile A, et al., Calorie restriction promotes mitochondrial biogenesis by inducing the expression of eNOS. 2005. *Science,* 310(5746): 314–7.

Nonomura S, Kanagawa H, Makimoto A. [Chemical constituents of polygonaceous plants. i. studies on the components of ko-j o-kon. (polygonum cuspidatum sieb.et zucc.)]. 1963. *Yakugaku Zasshi,* 83: 988–90. Japanese.

Opipari AW Jr, Tan L, Boitano AE, et al., Resveratrol-induced autophagocytosis in ovarian cancer cells. 2004. *Cancer Res*, 64(2): 696–703.

Phuah NH, Nagoor NH. Regulation of microRNAs by natural agents: New strategies in cancer therapies. 2014. *Biomed Res Int*, 2014: 804510.

Pintusorn H, Bhanuz D, Pathompong P, et al., Cytotoxic and antioxidant activities of Antidesma thwaitesianum Müll Arg (Euphorbiaceae) fruit and fruit waste extracts; 2015. *Trop J Pharm Res*, 14(4): 627–34.

Puangpronpitag D, Areejitranusorn P, Boonsiri P, et al., Antioxidant activities of polyphenolic compounds isolated from Antidesma thwaitesianum Müll. Arg. seeds and marcs. 2008. *J Food Sci,* 73(9): C648–53.

Puangpronpitag D, Yongvanit P, Boonsiri P, et al., Molecular mechanism underlying antiapoptotic and anti-inflammatory effects of Mamao (Antidesma thwaitesianum Mull. Arg.) polyphenolics in human breast epithelial cells. 2011. *Food Chem*, 127: 1450–8.

Pyo IS, Yun S, Yoon YE, et al., Mechanisms of aging and the preventive effects of resveratrol on age-related diseases. 2020. *Molecules,* 25(20): 4649. doi: 10.3390/molecules25204649. PMID: 33053864; PMCID: PMC7587336.

Ravikumar B, Sarkar S, Davies JE, et al., Regulation of mammalian autophagy in physiology and pathophysiology. 2010. *Physiol Rev*, 90(4): 1383–435.

Rege SD, Geetha T, Broderick TL, et al., Resveratrol protects β amyloid-induced oxidative damage and memory associated proteins in H19-7 hippocampal neuronal cells. 2015. *Curr Alzheimer Res,* 12(2): 147–56.

Ren B, Kwah MX, Liu C, et al., Resveratrol for cancer therapy: Challenges and future perspectives. 2021. *Cancer Lett,* 515: 63–72. doi: 10.1016/j.canlet.2021.05.001. Epub 2021 May 28. PMID: 34052324.

Savaskan E, Olivieri G, Meier F, et al., Red wine ingredient resveratrol protects from beta-amyloid neurotoxicity. 2003. *Gerontology*, 49(6): 380–3.

Savio M, Ferraresi A, Corpina C, et al., Resveratrol and its analogue 4,4'-Dihydroxy-trans-stilbene inhibit Lewis lung carcinoma growth in vivo through apoptosis, autophagy and modulation of the tumour microenvironment in a Murine model. 2022. *Biomedicines,* 10(8): 1784. doi: 10.3390/biomedicines10081784.

Selvaraj S, Sun Y, Sukumaran P, et al., Resveratrol activates autophagic cell death in prostate cancer cells via downregulation of STIM1 and the mTOR pathway. 2015. *Mol Carcinog*, 55(5): 818–31.

Seyed MA, Jantan I, Bukhari SN, et al., A comprehensive review on the chemotherapeutic potential of Piceatannol for cancer treatment, with mechanistic insights. 2016. *J Agric Food Chem,* 64(4): 725–37.

Singh V, Sharma P, Capalash N. DNA methyltransferase-1 inhibitors as epigenetic therapy for cancer. 2013. *Curr Cancer Drug Targets,* 13(4): 379–99.

Singh AP, Singh R, Verma SS, et al., Health benefits of resveratrol: Evidence from clinical studies. 2019. *Med Res Rev,* 39(5): 1851–91. doi: 10.1002/med.21565.

Slane BG, Aykin-Burns N, Smith BJ, et al., Mutation of succinate dehydrogenase subunit C results in increased O2.-, oxidative stress, and genomic instability. 2006. *Cancer Res,* 66(15): 7615–20.

Soeur J, Eilstein J, Léreaux G, et al., Skin resistance to oxidative stress induced by resveratrol: From Nrf2 activation to GSH biosynthesis. 2015. *Free Radic Biol Med,* 78: 213–23.

Soleas GJ, Diamandis EP, Goldberg DM. Resveratrol: A molecule whose time has come? And gone? 1997. *Clin Biochem,* (2): 91–113.

Song L, Chen L, Zhang X, et al., Resveratrol ameliorates motor neuron degeneration and improves survival in SOD1(G93A) mouse model of amyotrophic lateral sclerosis. 2014. *Biomed Res Int*, 2014: 483501.

St John SE, Jensen KC, Kang S, et al., Design, synthesis, biological and structural evaluation of functionalized resveratrol analogues as inhibitors of quinone reductase 2. 2013. *Bioorg Med Chem*, 21(19): 6022–37.

Stefanska B, Karlic H, Varga F, et al., Epigenetic mechanisms in anticancer actions of bioactive food components - the implications in cancer prevention. 2012. *Br J Pharmacol*, 167(2): 279–97.

Stivala LA, Savio M, Carafoli F, et al., Specific structural determinants are responsible for the antioxidant activity and the cell cycle effects of resveratrol. 2001. *J Biol Chem*, 276(25): 22586–94.

Sui T, Ma L, Bai X, et al., Resveratrol inhibits the phosphatidylinositide 3-kinase/protein kinase B/mammalian target of rapamycin signaling pathway in the human chronic myeloid leukemia K562 cell line. 2014. *Oncol Lett*, 7(6): 2093–8.

Tan L, Wang W, He G, et al., Resveratrol inhibits ovarian tumor growth in an in vivo mouse model. 2016. *Cancer*. 122(5): 722–9. doi: 10.1002/cncr.29793. Epub 2015 Nov 30. PMID: 26619367.

Tellone E, Galtieri A, Russo A, et al., Resveratrol: A focus on several neurodegenerative diseases. 2015. *Oxid Med Cell Longev*, 2015: 392169.

Terzibasi E, Valenzano DR, Cellerino A. The short-lived fish Nothobranchius furzeri as a new model system for aging studies. 2007. *Exp Gerontol*, 42(1–2): 81–9.

Tessitore L, Davit A, Sarotto I, et al., Resveratrol depresses the growth of colorectal aberrant crypt foci by affecting bax and p21(CIP) expression. 2000. *Carcinogenesis*, 21(8): 1619–22.

Tian B, Liu J. Resveratrol: A review of plant sources, synthesis, stability, modification and food application. 2020. *J Sci Food Agric*, 100(4): 1392–404. doi: 10.1002/jsfa.10152. Epub 2019 Dec 14. PMID: 31756276.

Thongchot S, Ferraresi A, Vidoni C, et al., Preclinical evidence for preventive and curative effects of resveratrol on xenograft cholangiocarcinogenesis. 2024. *Cancer Lett*, 582: 216589. doi: 10.1016/j.canlet.2023.216589.

Trincheri NF, Follo C, Nicotra G, et al., Resveratrol-induced apoptosis depends on the lipid kinase activity of Vps34 and on the formation of autophagolysosomes. 2008. *Carcinogenesis*, 29(2): 381–9.

Trincheri NF, Nicotra G, Follo C, et al., Resveratrol induces cell death in colorectal cancer cells by a novel pathway involving lysosomal cathepsin D. 2007. *Carcinogenesis*, 28(5): 922–31.

Turner RS, Thomas RG, Craft S, et al., Alzheimer's Disease Cooperative Study. A randomized, double-blind, placebo-controlled trial of resveratrol for Alzheimer disease. 2015. *Neurology*, 85(16): 1383–91.

Vahid F, Zand H, Nosrat-Mirshekarlou E, et al., The role dietary of bioactive compounds on the regulation of histone acetylases and deacetylases: A review. 2015. *Gen*, 562(1): 8–15.

Vallino L, Ferraresi A, Vidoni C, et al., Modulation of non-coding RNAs by resveratrol in ovarian cancer cells: In silico analysis and literature review of the anticancer pathways involved. 2020. *J Tradit Complement Med*, 10(3): 217–29. doi: 10.1016/j.jtcme.2020.02.006. PMID: 32670816; PMCID: PMC7340874.

van Ginkel PR, Sareen D, Subramanian L, et al., Resveratrol inhibits tumor growth of human neuroblastoma and mediates apoptosis by directly targeting mitochondria. 2007. *Clin Cancer Res*, 13(17): 5162–9.

Venkatadri R, Muni T, Iyer AK, et al., Role of apoptosis-related miRNAs in resveratrol-induced breast cancer cell death. 2016. *Cell Death Dis,* 7: e2104.

Vidoni C, Secomandi E, Castiglioni A, et al., Resveratrol protects neuronal-like cells expressing mutant Huntingtin from dopamine toxicity by rescuing ATG4-mediated autophagosome formation. 2018. *Neurochem Int,* 117: 174–87. doi: 10.1016/j.neuint.2017.05.013. Epub 2017 May 19. PMID: 28532681.

Xia N, Förstermann U, Li H. Resveratrol and endothelial nitric oxide. 2014. *Molecules,* 19(10): 16102–21.

Xiong W, Yin A, Mao X, et al., Resveratrol suppresses human glioblastoma cell migration and invasion via activation of RhoA/ROCK signaling pathway. 2016. *Oncology Letters,* 11(1): 484–90.

Zhang LX, Li CX, Kakar MU, et al., Resveratrol (RV): A pharmacological review and call for further research. 2021. *Biomed Pharmacother,* 143: 112164. doi: 10.1016/j.biopha.2021.112164.

Zhong LX, Wu ML, Li H, et al., Efficacy and safety of intraperitoneally administered resveratrol against rat orthotopic ovarian cancers. 2019. *Cancer Manag Res,* 11: 6113–24. doi: 10.2147/CMAR.S206301. PMID: 31456648; PMCID: PMC6620774.

Zhou J, Li SX, Wang W, et al., Variations in the levels of mulberroside A, oxyresveratrol, and resveratrol in mulberries in different seasons and during growth. 2013. *Scientific World Journal,* 2013: 380692.

7 Thai Traditional and Medicinal Uses of *Zingiber officinale* Roscoe

Shu-qin Cao, James M. Brimson, and Tewin Tencomano

7.1 INTRODUCTION TO GINGER

Zingiber officinale Roscoe (*Z.officinale*) (Figure 7.1), commonly known as ginger, is an herbaceous perennial aromatic plant belonging to the family Zingiberaceae. Its cultivation began in Southern Asia, and the rhizome of ginger is widely used for various foods and beverages. Also, it has a long history of folk medicine use dating back 2,500 years in China and India. In Thailand, *Z. officinale* is the quintessential panacea in the traditional herbal pharmacopoeia. Ginger is used for digestive disorders, including indigestion, sluggish digestion, colic, gastritis, diarrhoea, nausea, and vomiting (Zick et al. 2008). Also, ginger is known for treating colds, fevers, headaches, sore throats, bronchitis, asthma, gout, cramps, improving blood circulation, and menstruation (Ali et al. 2008). Based on the abundant research results, the bioactive phenolic compounds of ginger, such as gingerols, paradols, shogaols, and gingerones (Shukla and Singh 2007), have shown antioxidant (Shobana and Naidu 2000), anticancer (Katiyar et al. 1996), anti-inflammatory (Surh, 2002), anti-angiogenesis (Kim et al. 2005) and anti-atherosclerotic (Fuhrman et al. 2000) potentials.

7.2 PHYTOCHEMISTRY OF GINGER

Ginger could be listed as a hot herb with 477 constituents, such as ascorbic acid, capsaicin, beta-sitosterol, beta-carotene, curcumin, limonene, and tryptophan. Ginger is a rich source of volatile oil (1–3% of the weight of fresh ginger). Zingiberol, zingiberene, phellandrene, and linalool are significant oil constituents. They account for the aroma of the drug. It also contains many bioactive phenolic compounds, including non-volatile pungent substances such as gingerols, shogaols, and paradol (Shukla et al. 2007). Gingerol is the most abundant phytochemical in fresh ginger, and its

DOI: 10.1201/9781003508724-7

FIGURE 7.1 Photograph *of Zingiber officinale* Roscoe rhizome.

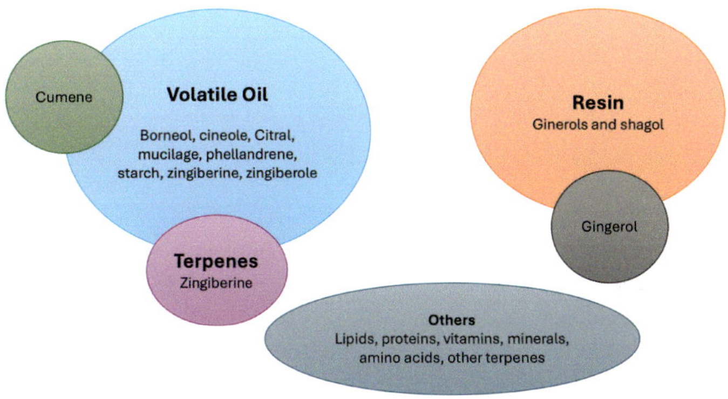

FIGURE 7.2 Volatile and non-volatile constituents of *Z. officinale*.

chemical homologs are differentiated by the length of their unbranched alkyl chains (Govindarajan 1983). Shogaols are dehydrated gingerol products that rarely exist in fresh ginger but are present in more significant amounts in dried ginger (Jolad et al. 2005). Paradol is another bioactive compound formed from the hydrogenation of shogaol. In addition, ginger contains a unique group of compounds called diarylheptanoids, including gingerones. The standardisation of the drug is based on the presence of pungent principles of the plant. The identified compounds of ginger are shown in Figure 7.2, and the most efficient constituents with their chemical structures are shown in Table 7.1.

TABLE 7.1
Natural Bioactive Compounds Identified in *Z. officinale*

Name	Structure	Examples of Potential Uses in Medicine
Zingiberene	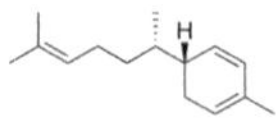	Antioxidant activity (Höferl et al. 2015) Antinociceptive activity (Jeena et al. 2013)
Zingerone	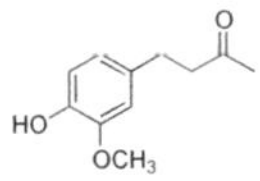	Anti-inflammatory and anticancer potential (Shukla et al. 2007)
Phellandrene (α- and β-phellandrene)	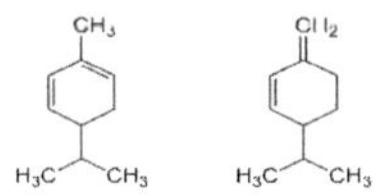	Anti-spasmodic effect (at the rat ileum) (Riyazi et al. 2007)
α-Bisabolene β-Bisabolene γ-Bisabolene	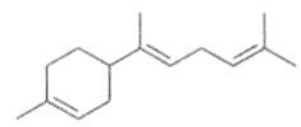 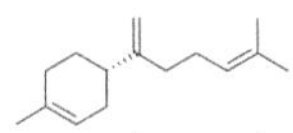 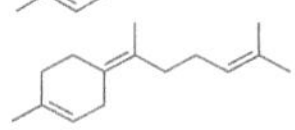	Anti-emetic activity (Ali et al. 2008) (Bisabolene only exists in young ginger.)
	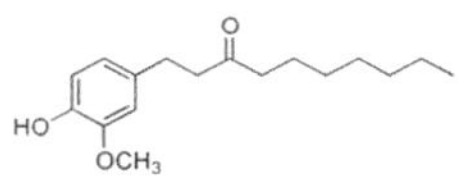	
	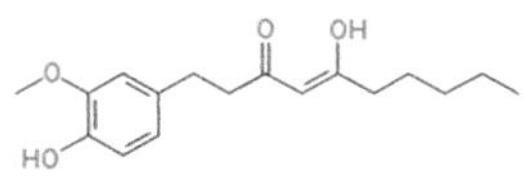	
[6]-Paradol	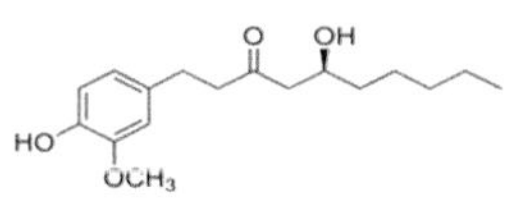	Antioxidant and anticancer properties in mouse model (Surh, Y. J et al. 1998)
[6]-gingerdione	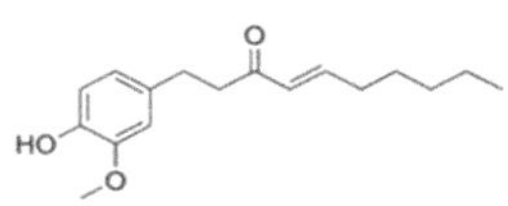	Antitumour (Hsu et al. 2005)
Gingerol	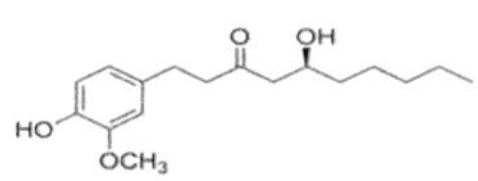 **n=4 [6]-gingerol** **n=2 [4]-gingerol** **n=5 [7]-gingerol** **n=6 [8]-gingerol** **n=8 [10]-gingerol**	Anti-tumour effect (Surh et al. 1998) Anti-ulcerative potential (Minaiyan et al. 2006)
Shogaol	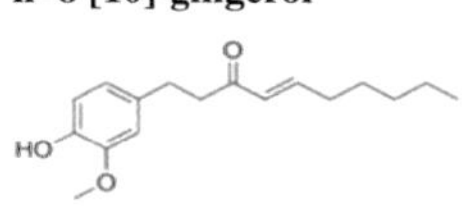 **n=4 [6]-shogaol** **n=2 [4]-shogaol** **n=6 [8]-shogaol** **n=8 [10]-shogaol** **n=10 [12]-shogaol**	Anticancer activities (Ling et al. 2010) Neuroprotective effect (Ha et al. 2012. Shim and Kwon 2012)

7.3 PHARMACOLOGICAL ACTIVITIES OF PLANT EXTRACTS AND CHEMICAL CONSTITUENTS

7.3.1 Antioxidant and Anti-inflammatory Activity

Oxidative stress is caused by an imbalance between the production of reactive oxygen and nitrogen species (ROS and RNS) and the biological system's ability to detoxify their potentially harmful effects through neutralisation by antioxidants. Oxidative stress damages biopolymers such as nucleic acids, polyunsaturated fatty acids, and proteins (Badarinath et al. 2010). It plays a significant role in an abundance of health problems such as inflammatory diseases, cancer, immune suppression, diabetes, ageing, and neurodegenerative disorders (Beal 1995; Poulsen et al. 1998). Therefore, antioxidants clear up these reactive species by scavenging the free radicals in a biological system and reducing the risk of complications. Inflammation is also demonstrated to be closely associated with various diseases and carcinogenesis, especially in the promotion stage. Thus, natural herbal products with antioxidant and anti-inflammatory potentials are expected to contribute to antitumour-promoting effects.

Ginger, with significant antioxidant and anti-inflammatory properties, has been revealed in previously published results. Like other natural plants used as antioxidant agents, ginger and its extracts contain several polyphenolic compounds due to their capacity to be donors of hydrogen atoms or electrons to capture free radicals and reduce the risk of oxidant stress and complications. A report from Masuda et al. (2004) identified the structures of more than 50 antioxidants isolated from the rhizomes of ginger. An *in vitro* DPPH assay shows that [6]-gingerol and its related compounds have similar kinetic behaviour without significant activity differences among the compounds with different alkyl chain lengths. Chung et al. (2001) reported that the naturally occurring phenolic compounds [6]-paradol and [6]-dehydroparadol in ginger were identified with anti-promotion effects with marked anti-inflammatory activity, which is as efficient as curcumin (one of the extensively investigated chemopreventive agents). *In vivo* tests showed that these compounds inhibited TPA-induced inflammation in mouse skin, and the vanillin (4-hydroxy-3-methoxyphenyl) moiety and the ketone functional group in paradols and hydroparadols were viewed as the significant factors for anti-inflammatory activity.

7.3.2 Anti-Tumourigenic and Anticancer Potential

A tumour, also known as a neoplasm, is an abnormal tissue mass. However, not all tumours are cancers. Tumours can be benign (not cancerous), pre-malignant (precancerous), or malignant (cancerous). The malignant tumour or malignant neoplasm-induced cancers are a group of fatal diseases initiated by the unchecked proliferation and abnormal growth of cells with the potential to invade or metastasise to other parts of the body. Cancer is a significant cause of morbidity and mortality across the world. About 14 million cancer cases and 8.2 million cancer-related deaths are estimated to have occurred in 2012 worldwide, and it's expected that cancer cases will rapidly rise by approximately 70% over the next two decades (Stewart and Wild

Riyazi, A., Hensel, A., Bauer, K., et al., The effect of the volatile oil from ginger rhizomes (Zingiber officinale), its fractions and isolated compounds on the 5-HT3 receptor complex and the serotoninergic system of the rat ileum. 2007. *Planta Medica*, 73(4): 355–362.

San Chang, J., Wang, K. C., Yeh, C. F., et al., Fresh ginger (Zingiber officinale) has antiviral activity against human respiratory syncytial virus in human respiratory tract cell lines. 2013. *Journal of Ethnopharmacology*, 145(1): 146–151.

Selkoe, D. J. Alzheimer's disease is a synaptic failure. 2002. *Science*, 298(5594): 789–791.

Seubert, P., Oltersdorf, T., Lee, M. G., et al., Secretion of β-amyloid precursor protein cleaved at the amino terminus of the β-amyloid peptide. 1993. *Nature*, 361(6409): 260–263.

Shim, S., & Kwon, J. Effects of [6]-shogaol on cholinergic signaling in HT22 cells following neuronal damage induced by hydrogen peroxide. 2012. *Food and Chemical Toxicology*, 50(5): 1454–1459.

Shobana, S., & Naidu, K. A. Antioxidant activity of selected Indian spices. 2000. *Prostaglandins, Leukotrienes and Essential Fatty Acids*, 62(2): 107–110.

Shukla, Y., & Singh, M. Cancer preventive properties of ginger: A brief review. 2007. *Food and Chemical Toxicology*, 45(5): 683–690.

Stewart, B. W. Wild CPe: World Cancer Report 2014. International Agency for Research on Cancer: Lyon, France, 2014.

Surh, Y. J. Antitumor promoting potential of selected spice ingredients with antioxidative and anti-inflammatory activities: A short review. 2002. *Food and Chemical Toxicology*, 40(8): 1091–1097.

Surh, Y. J., Park, K. K., Chun, K. S., et al., Anti-tumor-promoting activities of selected pungent phenolic substances present in ginger. 1998. *Journal of Environmental Pathology, Toxicology and Oncology: Official Organ of the International Society for Environmental Toxicology and Cancer*, 18(2): 131–139.

Vassallo, N. Natural compounds as therapeutic agents for amyloidogenic diseases. 2015. *Advances in Experimental Medicine and Biology*. (Vol. 863). Springer.

Wang, Q, Sun, AY, Simonyi, A, et al., Neuroprotective mechanisms of curcumin against cerebral ischemia-induced neuronal apoptosis and behavioral deficits. 2005. *Journal of neuroscience research*, 82(1): 138–48.

Yao, X., Yan, L., Yao, L. Acrylamide exposure impairs blood-cerebrospinal fluid barrier function. 2014. *Neural Regeneration Research,* 9(5): 555.

Zadorozhna, M., & Domenica M. Mechanisms of chemopreventive and therapeutic proprieties of ginger extracts in cancer. 2021. *International Journal of Molecular Sciences,* 22(12): 6599.

Zeng, G. F., Zong, S. H., Zhang, Z. Y., et al., The role of 6-gingerol on inhibiting Amyloid β-protein-induced apoptosis in PC12 cells. 2015. *Rejuvenation Research*, 18(5): 413–421.

Zhuang, X., Deng, Z. B., Mu, J., et al., Ginger-derived nanoparticles protect against alcohol-induced liver damage. 2015. *Journal of Extracellular Vesicles*, 4(1): 28713.

Zick, S. M., Djuric, Z., Ruffin, M. T., et al., Pharmacokinetics of 6-gingerol, 8-gingerol, 10-gingerol, and 6-shogaol and conjugate metabolites in healthy human subjects. 2008. *Cancer Epidemiology Biomarkers & Prevention,* 17(8): 1930–1936.

8 *Punica granatum* L. in Thai Traditional Medicine

Madhura B. de Silva, James M. Brimson, and Tewin Tencomnao

8.1 INTRODUCTION: THE POMEGRANATE

The pomegranate (Figure 8.1), *Punica granatum* L., is a large fruit-bearing deciduous shrub or small tree. It is called thabthim in Thai. The lesser-known Socotra pomegranate, *Punica protopunica*, was previously classified in the Punicaceae family. However, based on recent phylogenetic studies, the Angiosperm Phylogeny Group places Punica in the family Lythraceae. Native from the Himalayas in northern India to Iran, the pomegranate has been cultivated and naturalised over the entire Mediterranean region since antiquity (Jurenka, 2008). It had been grown in eastern India for thousands of years, and its ancient sources also linked Iran to Pakistan and China (Ismail et al., 2012). Pomegranate cultivation spread from the west of Iran to the Turkish European borders and to the American southwest, California, and Mexico (Celik et al., 2009; Lansky and Newman, 2007). The more arid regions of Southeast Asia, the East Indies, and tropical Africa also grow pomegranate (Jurenka, 2008).

The pomegranate symbolises life, longevity, health, femininity, fecundity, knowledge, morality, immortality, and spirituality, if not divinity (Mahdihassan, 1984). Considered "a pharmacy unto itself" in Ayurvedic medicine, the peels are deemed a powerful astringent and cure for diarrhoea and oral aphthae, the juice a "refrigerant" (Arseculeratne et al., 1985) and "blood tonic" (Lad and Frawley, 1986), and the bark and roots anthelmintic (Naovi et al., 1991). In Unani medicine of the Middle East and India (Izhar, 1989), pomegranate flowers are used as a medicament for diabetes mellitus (Saxena and Vikram, 2004). The peels are also regarded as a potent astringent and anti-inflammatory agent in traditional Chinese medicine. Besides disorders of the digestive tract, such as diarrhoea and dysentery, they are also used for traumatic bleeding, ulcers, and infections. Water decoction of dried pomegranate peels is used as a remedy for aphthae and ulcers in India, Tunisia, and Guatemala (Lansky and Newman, 2007). Most research on pomegranate has focused on its antioxidant, anti-inflammatory, and anti-cancer properties, which are three broad yet interconnected areas.

DOI: 10.1201/9781003508724-8

FIGURE 8.1 Photograph of *Punica granatum* L. fruit.

8.2 PHYTOCHEMISTRY OF POMEGRANATE

Pomegranate is rich in crude fibres, pectin, sugars, and tannins (Gil et al., 2000). The seed oil and juice contain flavonoids and anthocyanidins (Schubert et al., 1999), and the peel is enriched in ellagitannins, represented by ellagic acid (EA), gallic acid, and punicalagin (Ismail et al., 2012). Most of the therapeutic benefits from pomegranate consumption are derived, at least in part, from EA, ellagitannins (including punicalagins), punic acid, anthocyanins, flavonols, flavan-3-ols, and flavones (Table 8.1) (Gil et al., 2000; Kulkarni et al., 2007).

TABLE 8.1
Phytochemistry of *Punica granatum* L

Name	Structure	Examples of Potential Uses in Medicine
Ellagic acid	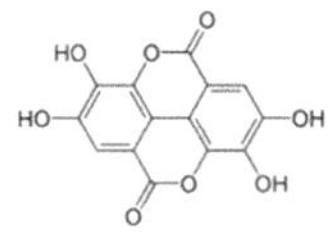	Anti-inflammatory activity (Panichayupakaranant et al., 2010; Mo et al., 2013, 2014) Antiproliferative, apoptotic, and antioxidant activities (Seeram et al., 2005) Anti-cancer activities (Naiki-Ito et al., 2015). Antibacterial activity (Panichayupakaranant et al., 2010)
Gallic acid	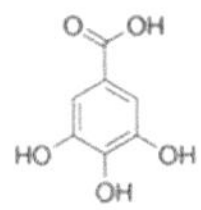	Anti-atherosclerotic activity (Khateeb et al., 2010; Rosenblat and Aviram, 2009)
Punicalagin	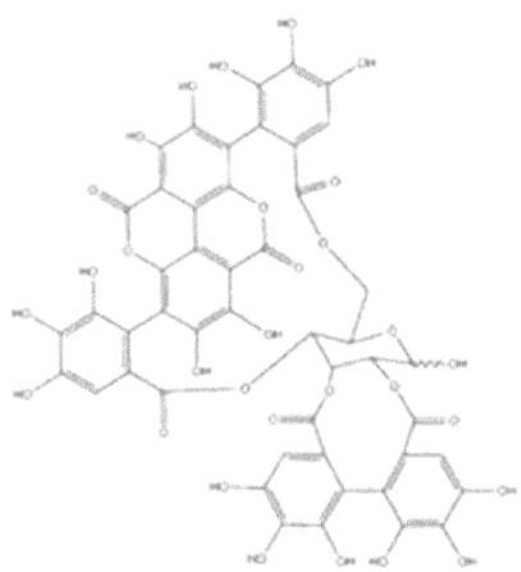	Antiproliferative, apoptotic, and antioxidant activities (Seeram et al., 2005) Anti-atherosclerotic activity (Khateeb et al., 2010; Rosenblat and Aviram, 2009)

8.3 MEDICINAL PROPERTIES OF POMEGRANATE EXTRACTS AND CHEMICAL CONSTITUENTS

8.3.1 Antioxidant Activity of Pomegranate

Pomegranate peel (PoP) and juice, which account for 92% of the antioxidant activity of the fruit, are abundant in phenolic compounds, including flavonoids (anthocyanins, catechins, and other complex flavonoids) and hydrolysable tannins (punicalin, pedunculagin, punicalagin, gallic, and ellagic acid) (Afaq et al., 2005; Negi et al., 2003; Zahin et al., 2010). Ferric-reducing antioxidant power (FRAP) assay has shown pomegranate peel extract (PoPx) to be the richest source of antioxidants among peel extracts of the most common fruits and that PoP demonstrated 2.8-fold higher antioxidant activity than pomegranate seed and leaf extracts (Guo et al., 2003; Okonogi et al., 2007; Negi et al., 2003; Tehranifar et al., 2011). PoP phenolics scavenge free radicals by electron donation that converts them to relatively more stable compounds, and the antioxidant power of pomegranate peel extract (PoPx) increases linearly with the concentration of phenolics up to the level of 400 μg/g (Naveena et al., 2008; Negi and Jayaprakasha, 2003).

Several studies have substantiated the chemoprotective role of the POM antioxidant fraction, suggesting an inhibitory and protective role against oxidative damage in carcinogenesis and its progression (Ismail et al., 2012). The attenuation of oxidative stress by pomegranate ellagitannins has been verified *in vivo* in treating placental dysfunction (Ismail et al., 2012). Oxidative stress and apoptosis were reduced in placental tissues by prenatal ingestion of the metabolic fractions of ellagitannins in juice form, substantiating *in vitro* results (Chen et al., 2012; Celik et al., 2009).

8.3.2 Anti-Inflammatory Activity of Pomegranate

The *in vitro* anti-inflammatory activity of standardised pomegranate rind extract (SPRE) containing 13% w/w EA was evaluated by Panichayupakaranant et al. (2010). SPRE displayed a potent inhibitory effect on nitric oxide (NO) production in murine macrophage-like RAW 264.7 cells, with an IC50 of 10.7 μg/ml. Mo et al. also investigated the *in vivo* effect. In one study (2014), SPRE was used in parallel with EA (13% w/w) against phenol-induced mouse ear oedema, a mouse contact dermatitis model. The ear oedema was dose-dependently reduced with 79.12% and 73.63% respective maximal inhibition by topical application of SPRE (5, 2.5, and 1 mg/ear) and EA (0.65, 0.325, and 0.13 mg/ear, equivalent to its content in SPRE), which also decreased myeloperoxidase (MPO) activity in the mouse ear up to 69.68% and 68.79%, respectively. The reference drugs triamcinolone (0.1 mg/ear) and diclofenac (1 mg/ear) inhibited the oedema by 73.63% and 37.91% and decreased the MPO activity by 76.66% and 80.14%, respectively. Another study (2013), which also used SPRE (5%, 2.5%, and 1%, w/w) and the equivalent EA (0.65%, 0.325%, and 0.13%, w/w), found that croton oil–induced mouse ear oedema was reduced dose-dependently, with 86.30% and 80.82% maximal inhibition, respectively. In the carrageenan-induced rat paw oedema and the CFA-induced polyarthritis, while SPRE attenuated the inflammatory responses dose-dependently, the equivalent EA was effective only at 0.65% and 0.325% doses. In the rat punctuate mechanical hyperalgesia and mouse formalin tests, SPRE (5%) and EA (0.65%) showed significant topical analgesic activities. They concluded that, as an anti-inflammatory agent, SPRE was more active than EA and that inhibiting infiltration of leukocytes and modulating the pro-inflammatory cytokines IL-β and TNF-α are the mechanisms through which these effects of SPRE are achieved.

8.3.3 Anti-Cancer Activity of Pomegranate

Seeram et al. (2005) evaluated the *in vitro* antiproliferative, apoptotic, and antioxidant activities of punicalagin, EA, a standardised total pomegranate tannin (TPT) extract, and pomegranate juice (PJ). PJ showed the most significant antiproliferative activity against all cell lines tested, namely, human oral (KB, CAL27), colon (HT-29, HCT116, SW480, SW620), and prostate (RWPE-1, 22Rv1) tumour cells, by inhibiting proliferation from 30% to 100%. Apoptosis was induced in both HT-29 and HCT116 colon cells by EA, punicalagin, and TPT at 100 μg/ml but only in HT-29 colon cells by PJ. Antioxidant activity displayed the trend PJ > TPT > punicalagin > EA. The multifactorial effects and chemical synergy of the action of multiple compounds explain PJ's superior bioactivity compared to its purified polyphenols.

The effect of pomegranate fruit juice (PFJ) and EA on prostate cancer, both *in vivo* and *in vitro*, has been demonstrated by Naiki-Ito et al. (2015). Using the transgenic rat for adenocarcinoma of prostate (TRAP) model, they found that, in the lateral prostate, PFJ decreased the incidence of adenocarcinoma, and both EA and PFJ suppressed prostate carcinogenesis progression and induced apoptosis by caspase-3 activation. In the ventral prostate, EA treatment significantly reduced the level of lipid peroxidation. The human prostate cancer cell lines, LNCaP (androgen-dependent), PC-3, and DU145 (androgen-independent), were used for the *in vitro* experiments. EA inhibited cell proliferation of LNCaP but not of PC-3 and DU145. Apoptosis induction in LNCaP by EA increased the Bax/Bcl-2 ratio and caspase-3 activation. EA treatment resulted in an increase in cell cycle–related proteins p21WAF, p27Kip, cdk2, and cyclin E and a decrease in cyclin D1 and cdk1. Thus, EA may be the active component of PFJ that exerts the above effects in androgen-dependent prostate carcinogenesis.

Khan et al. investigated the effect of the pomegranate fruit extract (PFE) on aggressive breast cancer phenotypes *in vitro* (2009). The aqueous PFE, which revealed mainly ellagitannins and phenolic acids as active constituents, dose-dependently inhibited NF-kB–dependent reporter gene expression associated with proliferation, invasion, and motility, coincident with the suppression of RhoC and RhoA protein expression, suggesting a role in reducing the metastatic potential of aggressive breast cancer species.

8.3.4 Cardiovascular Protective Role of Pomegranate

PoPx can impede the progression of atherosclerosis with a significant decrease in arterial foam cell levels by inhibiting LDL oxidation. The pomegranate polyphenols punicalagin, gallic acid, and, to a lesser extent, ellagic acid reduce the risk of atherosclerosis development by dose-dependently increasing hepatocyte paraoxonase-1 expression and secretion (Khateeb et al., 2010; Rosenblat and Aviram, 2009). PoP ellagitannins (10–100 μM) have shown cardiovascular disease preventive effects *in vitro*, whereas *in vivo* cardioprotective effects of pomegranate ellagitannins were relatively less, suggesting lower bioavailability of the antioxidant fractions in the fruit (Larrosa et al., 2010). Hassanpour Fard et al. corroborated that in a rat model, the cardioprotective effects of PoMx (100 mg/kg) occur via a reduction in creatine kinase-MB, lactate dehydrogenase, and glutathione (2011). The positive impact of POM polyphenols has also been established by others who showed their capacity to inhibit oxidation-sensitive genes, nitrous oxide synthase expression (de Nigris et al., 2005), and to downregulate redox-sensitive *ELK-1* and *p-JUN* genes, and endothelial nitrous oxide expression under induced endothelial wall shear stress (Larrosa et al., 2006, 2010). As a dietary fibre source, pomegranate peel powder has been assessed to treat hypercholesterolemia and atherosclerosis. Hossin showed that, in hypercholesterolemic rats, dietary supplementation with peel powder at a concentration of 5, 10, and 15 g/100 g for four weeks significantly reduced serum total cholesterol, triglycerides, LDL, and lipid peroxidation levels (2009).

8.3.5 Neuroprotective Role of Pomegranate

A significant cause of illness and death in severely preterm, low birth-weight infants is neonatal hypoxic-ischemic (HI) brain injury (Huang and Castillo, 2008), which has been associated with an increase in reactive oxygen species (Gulcan et al., 2005). Holtzman's group used pregnant mice given PJ in drinking water or a control beverage, experimentally induced HI brain injury in the neonatal offspring, and discovered that neonates from the PJ group had significantly less brain tissue loss (64% decrease) and significantly decreased hippocampal caspase-3 activity (84% decrease) compared to neonates from the control group (Loren et al., 2005; West et al., 2007), suggesting an antioxidant-driven neuroprotective effect of PJ conferred from mother to neonate.

Hartman et al. have demonstrated the neuroprotective properties of pomegranate polyphenols using APP (SW)/Tg2576 transgenic mice with Alzheimer-like pathology. Compared to control mice given sugar water, those treated with PJ had 50% less accumulation of soluble amyloid-beta and less hippocampal amyloid deposition, and also showcased improved learning of water maze tasks and swam faster (Hartman et al., 2006).

8.3.6 Antibacterial Activity of Pomegranate

Using the disc diffusion and broth microdilution methods, Panichayupakaranant et al. (2010) determined the antibacterial activity of SPRE containing 13% w/w EA, which displayed a potent bacteriostatic effect against a gram-positive anaerobe, *Propionibacterium acnes*, with a minimum inhibitory concentration (MIC) of 15.6 μg/ml, and gram-positive facultative anaerobes, *Staphylococcus aureus* and *Staphylococcus epidermidis*, with MICs of 7.8–15.6 μg/ml.

The activity of medicinal plant extracts against hospital isolates of methicillin-resistant Staphylococcus aureus (MRSA) has been investigated by Voravuthikunchai and Kitpipit (2005), who found that the most effective were ethanolic extracts of *Garcinia mangostana*, *Punica granatum,* and *Quercus infectoria*, with MICs of 0.05–0.4, 0.2–0.4, and 0.2–0.4 mg/ml respectively, for MRSA isolates, and of 0.1, 0.2, and 0.1 mg/ml respectively, for *S. aureus* ATCC 25923. MBCs were 0.1–0.4, 1.6–3.2, and 0.4–1.6 mg/ml for MRSA isolates and 0.4, 3.2, and 1.6 mg/ml for *S. aureus* ATCC 25923.

Both aqueous and ethanolic extracts of *Quercus infectoria* and aqueous extract of *Punica granatum* have also been shown to be highly effective against enterohaemorrhagic *Escherichia coli* O157:H7 by Voravuthikunchai et al., with the best MIC and MBC values of 0.09, 0.78, and 0.19, 0.39 mg/ml, respectively (2004).

Mucoadhesive polyethyleneimine–dextran sulphate nanoparticles (PDNPs) containing *Punica granatum* peel extract (PGE) have been validated as a novel sustained-release antimicrobial by Tiyaboonchai et al. for oral malodor reduction and caries prevention (2015). *Ex vivo* wash-off test confirmed PDNPs to be an excellent mucoadhesive drug delivery system, and *in vitro* dissolution studies revealed a prolonged release characteristic with a burst release within 5 minutes for PGE-loaded PDNPs, which remained effective against oral bacteria. The broth macro dilution test was used to investigate the antimicrobial activity of free PGE and PGE-loaded PDNPs against periodontitis-causing bacteria *Streptococcus mutans*, *Streptococcus sanguinis*, and *Porphyromonas gingivalis*. Both free PGE and PGE-loaded PDNPs showed significant antibacterial activity against the test pathogens, with similar MIC values, verifying that PGE could be released from PDNPs in active amounts. Relative to the positive control chlorhexidine, PGE possessed a MIC value of approximately five times less and a low range of 31.2–62.5 μg/ml (Maddox et al., 2010), validating it as a potent antibacterial.

8.3.7 Wound-Healing Activity of Pomegranate

SPRE and its principal antioxidant constituent, EA (13%, w/w), have also been investigated for *in vivo* wound-healing potential in three rat dermal wound models by Mo et al. (2014a). In the incision wound model, both SPRE (5 and 2.5%) and its equivalent amount of EA (0.65 and 0.325%) increased the tensile strength by a maximum of 35.43 and 31.82%, respectively, whereas in the excision and burn wound models, SPRE accelerated wound contraction at 5 and 2.5%. In comparison, EA was effective only at 0.65%. Furthermore, SPRE enhanced collagen synthesis by a maximum of 21.83 mg/g and dose-dependently inhibited neutrophil infiltration. At the same time, EA had a milder inhibitory effect on neutrophil infiltration and had no effect on increasing collagen accumulation, indicating the superiority of SPRE to its marker compound EA in facilitating wound healing.

8.3.8 Anti-diabetic Activity of Pomegranate

The positive effect of pomegranate flower extract (PFLE) on lipid profiles and cardiac fibrosis of Zucker fatty diabetic rats has been illustrated by Huang et al. (2005a and 2005b).

The effect of 50 ml/day pomegranate juice (PJ) on oxidative stress, blood sugar, and lipid profiles in ten type 2 diabetic patients (history of diabetes for 4–10 years) and ten healthy controls (ages 35–71) was investigated for three months by Rosenblat et al. (2006). Relative to the controls, diabetic patients had 2.8 times higher triglyceride levels, 28% lower HDL cholesterol, and 59% higher haemoglobin A1C (HbA1C) values. At the same time, insulin was slightly lower, and C-peptide (a proinsulin metabolite marker for endogenously secreted insulin) was marginally higher at baseline (indicating slight hyperinsulinemia). PJ consumption for three months lowered serum C-peptide values by 23% in diabetic patients relative to baseline, suggesting improved insulin sensitivity but did not significantly affect triglyceride, HDL cholesterol, HbA1C, glucose, or insulin values. Oxidative stress was also significantly reduced in diabetic patients by PJ, as verified by a 56% reduction in lipid peroxides and a 28% reduction in TBARS compared to baseline serum levels. Furthermore, after PJ consumption, human monocyte-derived macrophages (an early development in foam cell formation and atherogenesis) displayed a 39% decrease in the uptake of oxidised LDL in diabetic patients. The conclusion was that PJ consumption significantly affected atherogenesis via reduced oxidative stress and did not adversely affect diabetic parameters despite the sugars naturally present in it.

8.4 CONCLUSIONS

Based on published information, pomegranate is a plausible candidate responsible for antioxidant, anti-inflammatory, anti-cancer, cardiovascular protective, neuroprotective, antibacterial, wound healing, and anti-diabetic effects. In addition to supporting established use in Thai and other traditional medicinal systems, these discoveries are significant for further drug development of well-supported pathological conditions.

REFERENCES

Afaq F., Saleem M., Krueger C.G., Reed J.D., Mukhtar H. (2005) Anthocyanin- and hydrolyzable tannin-rich pomegranate fruit extract modulates MAPK and NF-kappaB pathways and inhibits skin tumorigenesis in CD-1 mice. *Int J Cancer*, 113, 3, 423–433.

Arseculeratne S.N., Gunatilaka A.A.L., Panabokke R.G. (1985) Studies on medicinal plants of Sri Lanka. Part 14. Toxicity of some traditional medicinal herbs. *J Ethnopharmacol*, 13, 323–335.

Celik I., Temur A., Isik I. (2009) Hepatoprotective role and antioxidant capacity of pomegranate (Punica granatum) flowers infusion against trichloroacetic acid-exposed in rats. *Food Chem Toxicol*, 47, 1, 145–149.

Chen B., Tuuli M.G., Longtine M.S., Shin J.S., Lawrence R., Inder T., Michael Nelson D. (2012) Pomegranate juice and punicalagin attenuate oxidative stress and apoptosis in human placenta and in human placental trophoblasts. *Am J Physiol Endocrinol Metab*, 302, 9, 1142–1152.

de Nigris F., Williams-Ignarro S., Lerman L.O., Crimi E., Botti C., Mansueto G., D'Armiento F.P., DeRosa G., Sica V., Ignarro L.J., Napoli C. (2005) Beneficial effects of pomegranate juice on oxidation-sensitive genes and endothelial nitric oxide synthase activity at sites of perturbed shear stress. *PNAS*, 102, 13, 4896–4901.

Gil M.I., Toma's-Barbera'n F.A., Hess-Pierce B., Holcroft D.M., Kader A.A. (2000) Antioxidant activity of pomegranate juice and its relationship with phenolic composition and processing. *J Agric Food Chem*, 48, 10, 4581–4589.

Gulcan H., Ozturk I.C., Arslan S. (2005) Alterations in antioxidant enzyme activities in cerebrospinal fluid related with severity of hypoxic ischemic encephalopathy in newborns. *Biol Neonate*, 88, 87–91.

Guo C., Yang J., Wei J., Li Y., Xu J., Jiang Y. (2003) Antioxidant activities of peel, pulp, and seed fractions of common fruits as determined by FRAP assay. *Nutr Res*, 23, 1719–1726.

Hartman R.E., Shah A., Fagan A.M., Schwetye K.E., Parsadanian M., Schulman R.N., Finn M.B., Holtzman D.M. (2006) Pomegranate juice decreases amyloid load and improves behavior in a mouse model of Alzheimer's disease. *Neurobiol Dis*, 24, 3, 506–515.

Hassanpour Fard M., Ghule A.E., Bodhankar S.L., Dikshit M. (2011) Cardioprotective effect of whole fruit extract of pomegranate ondoxorubicin-induced toxicity in rat. *Pharm Biol*, 49, 4, 377–382.

Hossin F.L.A. (2009) Effect of Pomegranate (Punica granatum) peels and it's extract on obese hypercholesterolemic rats. *Pak J Nutr*, 8, 1251–1257.

Huang B.Y., Castillo M. (2008) Hypoxic-ischemic brain injury: Imaging findings from birth to adulthood. *Radiographics*, 28, 417–439.

Huang T.H., Peng G., Kota B.P., Li G.Q., Yamahara J., Roufogalis B.D., Li Y. (2005a) Pomegranate flower improves cardiac lipid metabolism in a diabetic rat model: Role of lowering circulating lipids. *Br J Pharmacol*, 145, 6, 767–774.

Huang T.H., Yang Q., Harada M., Li G.Q., Yamahara J., Roufogalis B.D., Li Y. (2005b) Pomegranate flower extract diminishes cardiac fibrosis in Zucker diabetic fatty rats: Modulation of cardiac endothelin-1 and nuclear factor-kappaB pathways. *J Cardiovasc Pharmacol*, 46, 6, 856–862.

Ismail T., Sestili P., Akhtar S. (2012) Pomegranate peel and fruit extracts: A review of potential anti-inflammatory and anti-infective effects. *J Ethnopharmacol*, 143, 2, 397–405.

Izhar N. (1989) The Unani traditional medical system in India: A case study in health behavior. *Geographia Medica*, 19, 163–185.

Jurenka J.S. (2008) Therapeutic applications of pomegranate (Punica granatum L.): A review. *Altern Med Rev*, 13, 2, 128–144.

Khan G.N., Gorin M.A., Rosenthal D., Pan Q., Bao L.W., Wu Z.F., Newman R.A., Pawlus A.D., Yang P., Lansky E.P., Merajver S.D. (2009) Pomegranate fruit extract impairs invasion and motility in human breast cancer. *Integr Cancer Ther*, 8, 3, 242–253.

Khateeb J., Gantman A., Kreitenberg A.J., Aviram M., Fuhrman B. (2010) Paraoxonase 1 (PON1) expression in hepatocytes is upregulated by pomegranate polyphenols: A role for PPAR-gamma pathway. *Atherosclerosis*, 208, 1, 119–125.

Kulkarni A.P., Mahal H.S., Kapoor S., Aradhya S.M. (2007) In vitro studies on the binding, antioxidant, and cytotoxic actions of punicalagin. *J Agric Food Chem*, 55, 4, 1491–1500.

Lad V., Frawley D. (1986) *The Yoga of Herbs*. 1st ed. Santa Fe: Lotus Press.

Lansky E.P., Newman R.A. (2007) Punica granatum (pomegranate) and its potential for prevention and treatment of inflammation and cancer. *J Ethnopharmacol*, 109, 2, 177–206.

Larrosa M., Garcia-Conesa M.T., Espin J.C., Tomas-Barberan F.A. (2010) Ellagitannins, ellagic acid and vascular health. *Mol Asp Med*, 31, 513–539.

Larrosa M., Gonzalez-Sarrias A., Garcia-Conesa M.T., Tomas-Barberan F.A., Espin J.C. (2006) Urolithins, ellagic acid-derived metabolites produced by human colonic microflora, exhibit estrogenic and antiestrogenic activities. *J Agric Food Chem*, 54, 1611–1620.

Loren D.J., Seeram N.P., Schulman R.N., Holtzman D.M. (2005) Maternal dietary supplementation with pomegranate juice is neuroprotective in an animal model of neonatal hypoxic-ischemic brain injury. *Pediatr Res*, 57, 858–864.

Maddox C.E., Laur L.M., Tian L. (2010) Antibacterial activity of phenolic compounds against the phytopathogen Xylella fastidiosa. *Curr Microbiol*, 60, 53–58.

Mahdihassan S. (1984) Outline of the beginnings of alchemy and its antecedents. *Am J Chin Med*, 12, 32–42.

Mo J., Panichayupakaranant P., Kaewnopparat N., Nitiruangjaras A., Reanmongkol W. (2013) Topical anti-inflammatory and analgesic activities of standardized pomegranate rind extract in comparison with its marker compound ellagic acid in vivo. *J Ethnopharmacol*, 148, 3, 901–908.

Mo J., Panichayupakaranant P., Kaewnopparat N., Nitiruangjaras A., Reanmongkol W. (2014a) Wound healing activities of standardized pomegranate rind extract and its major antioxidant ellagic acid in rat dermal wounds. *J Nat Med*, 68, 2, 377–386.

Naiki-Ito A., Chewonarin T., Tang M., Pitchakarn P., Kuno T., Ogawa K., Asamoto M., Shirai T., Takahashi S. (2015) Ellagic acid, a component of pomegranate fruit juice, suppresses androgen-dependent prostate carcinogenesis via induction of apoptosis. *Prostate*, 75, 2, 151–160.

Naovi S.A.H., Khan M.S.Y., Vohora S.B. (1991) Antibacterial, anti-fungal and anthelmintic investigations on Indian medicinal plants. *Fitoterapia*, 62, 221–228.

Naveena B.M., Sen A.R., Kingsly R.P., Singh D.B., Kondaiah N. (2008) Antioxidant activity of pomegranate rind powder extract in cooked chicken patties. *Int J Food Sci Technol*, 43, 1807.

Negi P.S., Jayaprakasha G.K. (2003) Antioxidant and antibacterial activities of Punica granatum peel extracts. *Food Microbiol Safety*, 68, 1473–1477.

Negi P.S., Jayaprakasha G.K., Jena B.S. (2003) Antioxidant and antimutagenic activities of pomegranate peel extracts. *Food Chem*, 80, 393–398.

Okonogi S., Duangrat C., Anuchpreeda S., Tachakittirungrod S., Chowwanapoon-pohn S. (2007) Comparison of antioxidant capacities and cytotoxicities of certain fruit peels. *Food Chem*, 103, 839–846.

Panichayupakaranant P., Tewtrakul S., Yuenyongsawad S. (2010) Antibacterial, anti-inflammatory and anti-allergic activities of standardized pomegranate rind extract. *Food Chem*, 123, 400–403.

Rosenblat M., Aviram M. (2009) Paraoxonases role in the prevention of cardiovascular diseases. *Biofactors*, 35, 1, 98–104.

Rosenblat M., Hayek T., Aviram M. (2006) Anti-oxidative effects of pomegranate juice (PJ) consumption by diabetic patients on serum and on macrophages. *Atherosclerosis*, 187, 363–371.

Saxena A., Vikram N.K. (2004) Role of selected Indian plants in management of type 2 diabetes: A review. *J Altern Complement Med*, 10, 369–378.

Schubert S.Y., Lansky E.P., Neeman I. (1999) Antioxidant and eicosanoid enzyme inhibition properties of pomegranate seed oil and fermented juice flavonoids. *J Ethnopharmacol*, 66, 1, 11–17.

Seeram N.P., Adams L.S., Henning S.M., Niu Y., Zhang Y., Nair M.G., Heber D. (2005) In vitro antiproliferative, apoptotic and antioxidant activities of punicalagin, ellagic acid and a total pomegranate tannin extract are enhanced in combination with other polyphenols as found in pomegranate juice. *J Nutr Biochem*, 16, 6, 360–367.

Tehranifar A., Selahvarzi Y., Kharrazi M., Bakhsh V. (2011) High potential of agro-industrial by-products of pomegranate (Punica granatum L.) as the powerful antifungal and antioxidant substances. *Ind Crops Prod*, 34, 1523–1527.

Tiyaboonchai W., Rodleang I., Ounaroon A. (2015) Mucoadhesive polyethylenimine-dextran sulfate nanoparticles containing Punica granatum peel extract as a novel sustained-release antimicrobial. *Pharm Dev Technol*, 20, 4, 426–432.

Voravuthikunchai S.P., Kitpipit L. (2005) Activity of medicinal plant extracts against hospital isolates of methicillin-resistant Staphylococcus aureus. *Clin Microbiol Infect*, 11, 6, 510–512.

Voravuthikunchai S., Lortheeranuwat A., Jeeju W., Sririrak T., Phongpaichit S., Supawita T. (2004) Effective medicinal plants against enterohaemorrhagic Escherichia coli O157:H7. *J Ethnopharmacol*, 94, 1, 49–54.

West T., Atzeva M., Holtzman D.M. (2007) Pomegranate polyphenols and resveratrol protect the neonatal brain against hypoxic-ischemic injury. *Dev Neurosci*, 29, 363–372.

Zahin M., Aqil F., Ahmad I. (2010) Broad spectrum antimutagenic activity of antioxidant active fraction of punica granatum L. peel extracts. *Mutat Res*, 703, 2, 99–107.

9 The Memory-Enhancing Effects of Thai Medicinal Herb *Bacopa monnieri*

James M. Brimson

9.1 INTRODUCTION

Bacopa monnieri (L) Wettst is a medicinal herb found in Thailand. It is a member of the Scrophulariaceae family (Uabundit et al. 2010) and is usually found growing on the banks of rivers and lakes. It is a small shrub with thick oval leaves and white flowers (Figure 9.1). The common name in India and Thailand is "Bharmi" or "Bhumi." The herb is highly prized among local cultures for its medicinal properties, and with its relatively slow growth and high demand, it has become one of the most adulterated herbs on the market; as such, there have been efforts to standardise the culture of this plant in the laboratory setting (Banerjee and Sarika 2008).

B. monnieri is known for its content of a group of compounds known collectively as Bacoside A, which consists of Bacoside A3, Bacopaside II, Bacopaside X, and Bacopasaponin C 5 (Table 9.1). This mix of compounds is used as a standard for confirming the authenticity of a *B. monnieri* extract (Figure 9.2) (Sivaramakrishna et al. 2005).

9.2 MEDICINAL PROPERTIES OF *B. MONNIERI* EXTRACTS AND ITS CHEMICAL CONSTITUENTS

9.2.1 Memory-Enhancing Effects of *B. monnieri*

B. monnieri has long been used for its memory-enhancing effects in Ayurvedic medicine (Gupta and Kale 2023). There has been much research into *B. monnieri's* cognitive enhancing abilities; in animal studies, olfactory bulbectomised mice performed better in the Y maze and fear conditioning tests when treated with *B. monnieri* (Le et al. 2013). Furthermore, the amnesic effects of scopolamine in the Morris water maze test can be reversed with *B. monnieri* treatment (Saraf, Anand, and Prabhakar 2010). These animal studies have translated into positive results in human trials, with several clinical trials investigating the behavioural and memory-enhancing effects of herb formulations containing *B. monnieri* (Ojha, Kumar,

DOI: 10.1201/9781003508724-9

FIGURE 9.1 *Bacopa monnieri* showing the small ovate leaves and white flowers.

and Rai 2007; Katz et al. 2010; Upadhyay et al. 2002; D'souza 1991; Patel 1991; Quadri 1993; Nakazawa et al. 2008; Dave, Chauvan, and Dalvi 1993). Many of these studies used an Ayurvedic formulation of Mentat® (Himalaya®), a popular herbal supplement available widely throughout Thailand and Southeast Asia. Each tablet is reported to contain the following herbs: *B. monnieri* (136 mg), *Centella asiatica* (70 mg), *Withania somnifera* (52 mg), *Evolvulus alsinoides* (52 mg), *Nardostachys jatamansi* (52 mg), *Valeriana wallichii* (50 mg), *Embelia ribes* (50 mg), *Prunus amygdalus* (50 mg), *Tinospora cordifolia* (36 mg), *Terminalia chebula* (36 mg), *Emblica officinalis* (36 mg), *Oroxylum indicum* (32 mg), and *Celastrus paniculatus* (32 mg) (Kalra and Kulkarni 2002).

From the clinical trials cited above, there were few, if any, side effects, and in Mentat® several other herbs in the Mentat® formulation, *B. monnieri* is the main herb, and as described earlier, it is effective in improving memory in animal models. BacoMind®, a herbal remedy that is safe in phase I clinical trials (Kean et al. 2015), consists of compounds found in *B.* monnieri rather than a mixture of herbs (bacopaside I (>7.0% w/w), jujubogenin isomer of bacopasaponin C (>7.0% w/w), bacopaside II (>5.5% w/w), bacoside A3 (>5.0% w/w), bacopasaponin C (>4.5% w/w), bacosine (>1.5% w/w), β-sitosterol-D-glucoside (>0.3% w/w), luteolin (>0.2% w/w), and apigenin (>0.1% w/w), and it has been found to give a statistically significant increase in memory of children with ADHD (Usha et al. 2008).

TABLE 9.1
Phytochemical Constituents of *B. monnieri*

Name	Structure
Bacoside A3	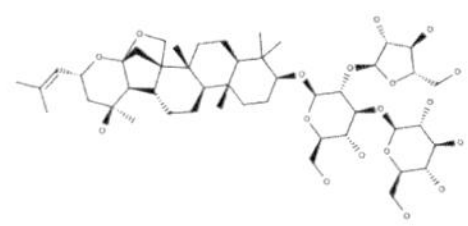
Bacomoside A	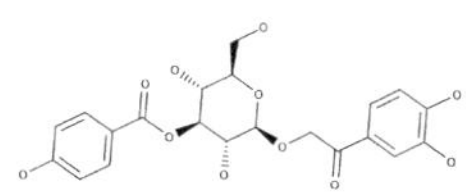
Bacopaside II	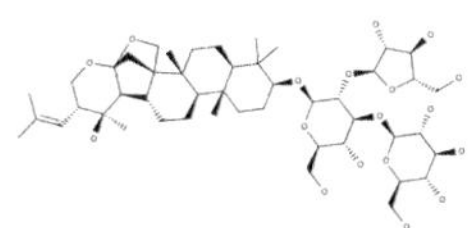
Bacopaside VII	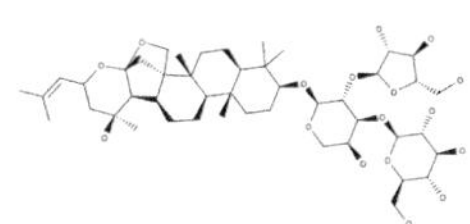
Bacopaside X	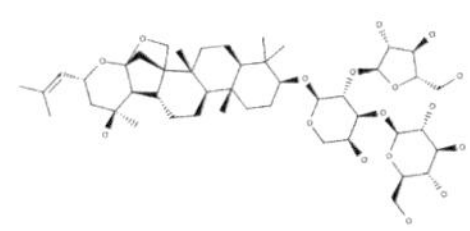
Bacomosaponin A	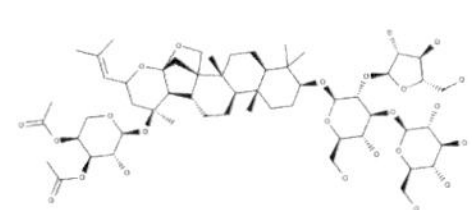
Bacopasaponin C	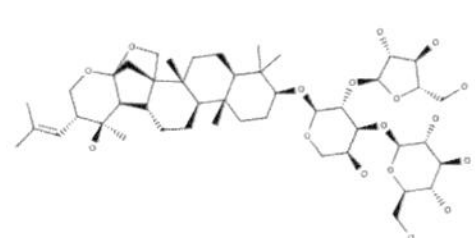
Bacopasaponin E	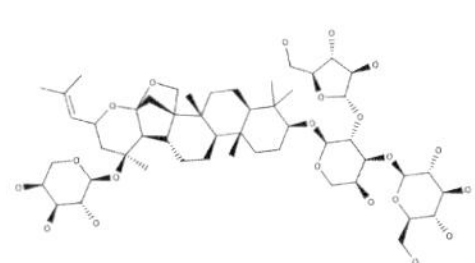
Bacopasaponin F	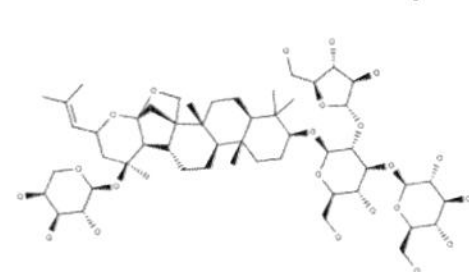

(Continued)

TABLE 9.1 CONTINUED
Phytochemical Constituents of *B. monnieri*

Name	Structure
Cucurbitacin E	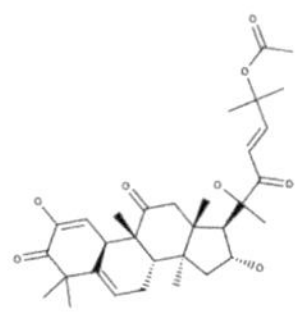
Cucurbitacin I	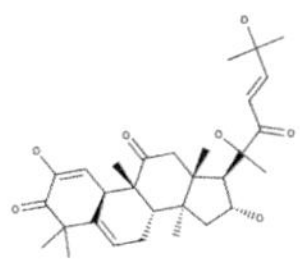
monnieraside III	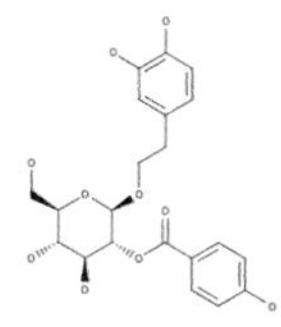
Plantainoside B	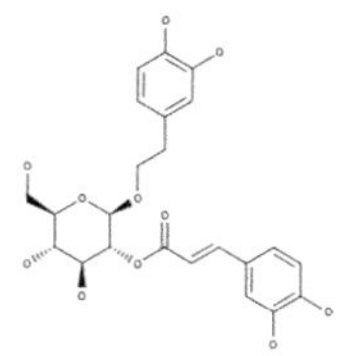

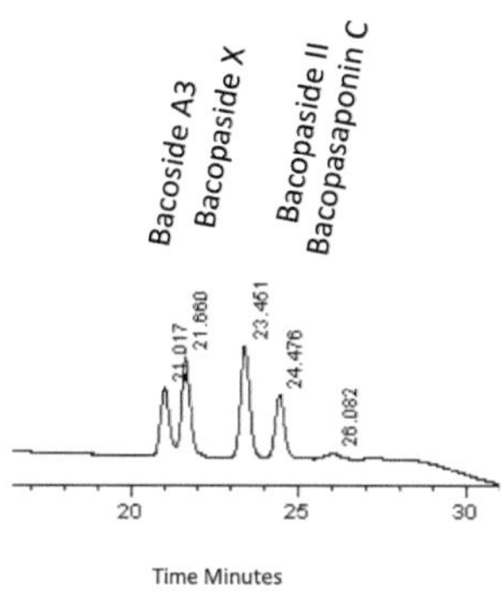

FIGURE 9.2 HPLC fingerprint of *B.monnieri*.

9.2.2 Alzheimer's Disease Treatment

B. monnieri's effectiveness in improving memory has not gone unnoticed in Alzheimer's research. *B. monnieri* has been shown *in vitro* to protect primary cultured neurons from the detrimental effects of β amyloid, the protein fragment crucially associated with Alzheimer's disease (Limpeanchob et al. 2008). In animal models of dementia, *B. monnieri* has proved effective, such as in mice overexpressing the amyloid precursor protein APP and ethylcholine aziridinium ion treated mice (Uabundit et al. 2010; Saini, Singh, and Sandhir 2012). Furthermore, double-blind clinical trials in Alzheimer's patients and patients with age-associated memory loss have been conducted with some success (Goswami et al. 2011; Amit et al. 2009).

Recently, it has been shown that *B. monnieri* hexane extracts exhibit significant neuroprotective effects, preventing the detrimental effects of glutamate in HT-22 cells. Furthermore, *B. monnieri* has lifespan-extending properties in *C. elegans* wild-type strains (Brimson 2020). A recent study showed that *B. monnieri* could protect against amyloid-beta toxicity in differentiated PC-12 cells and provide lifespan-extending effects in amyloid-beta overexpressing strains of *C. elegans*. Moreover, these effects appeared sensitive to sigma-1 receptor antagonist action, suggesting that the phytochemicals found in *B. monnieri* have activity at the sigma-1 receptor (Brimson et al. 2022).

The Goswami et al. study used a standardised extract of *B. monnieri* (Bacognize®), which contained Bacoside A3 4.6% w/w, Bacopaside II 2.1% w/w, Bacopaside X 1.7% w/w, and Bacopasaponin C 3.0% w/w. Their study showed that six months of treatment with the *B. monnieri* extracts gave statistically significant improvements in orientation, attention, and language abilities. They also reported an anecdotal improvement in the patient's well-being. This study, however, like many of the others with *B. monnieri* concerning Alzheimer's disease, does not include a positive control, i.e., they do not compare *B. monnieri* extract to an existing drug such as donepezil. Without such a study, it is hard to put into context the effectiveness of the extract as a potential drug for Alzheimer's disease. Although the results appear promising, until a double-blind placebo-controlled study compares *B. monnieri* to donepezil, we will not truly know the extract's efficacy in Alzheimer's treatment. A recent meta-analysis study has suggested no evidence that *B. monnieri* has any memory improvement in healthy individuals. Still, there may be some benefit for those suffering from memory impairment, but more clinical trials using *B. monnieri* are required (Brimson 2021).

9.2.3 Neuroprotective Mechanism

A study using PSAPP-expressing mice showed that *B. monnieri* could reduce the amount of beta-amyloid found in different brain regions in short- and long-term treatments (Hitt et al. 2017). There have been several studies linking cerebral blood flow to the deposition of beta-amyloid (Hooijmans et al. 2007). Furthermore, studies have shown that *B. monnieri* treatment results in an increase in cerebral blood flow. The various studies involving *B. monnieri* and Alzheimer's indicate ROS mitigation

10 Traditional Thai Medicinal Uses of *Curcuma comosa* Roxb.

Madhura B. de Silva, James M. Brimson, and Tewin Tencomnao

10.1 INTRODUCTION TO *CURCUMA COMOSA* ROXB.

Curcuma comosa Roxb. (Figure 10.1) is a flowering plant in the ginger family, Zingiberaceae. It is widely distributed in tropical and subtropical regions of Asia, especially Thailand, Indonesia, and Malaysia (Su et al. 2010), where it is a typical traditional herbal remedy for gynaecological diseases due to its oestrogenic properties. Known as "*waan chak mod look*" in Thai, it has traditionally been used for treating uterine disorders and ovarian hormone deficiency (Piyachaturawat et al. 1995) and as a dietary supplement for relief from various unpleasant peri- and postmenopausal symptoms in indigenous medicine of Thailand (Prasannarong et al. 2012). The rhizome of *C. comosa* has long been used as an anti-inflammatory to alleviate uterine inflammation, post-partum uterine bleeding, and peri-menopausal bleeding (Thampithak et al. 2009). Other uses include an aromatic stomachic and a cholagogue (Piyachaturawat et al. 1999).

10.2 PHYTOCHEMISTRY

Piyachaturawat et al. demonstrated that the hexane extract of *C. comosa* rhizomes displayed typical characteristics of oestrogenic action in rats, such as uterotrophic activity and inducing cornification of the vaginal epithelium and keratinisation of the vaginal mucosal surface, and attributed this to the presence of phytoestrogens in the hexane extract (1995). Several pure compounds isolated from the rhizomes of *C. comosa*, two groups of structures, sesquiterpenes and diarylheptanoids, have been identified (Table 10.1) (Suksamrarn et al. 2008, Xu et al. 2008). Oestrogenic activity is comparable to oestradiol and genistein, as shown by some of the isolated diarylheptanoids and their modified analogues (Suksamrarn et al., 2008). The two active pure diarylheptanoids from *C. comosa* most examined for specific influence are ASPP 049 and ASPP 092 (Vattanarongkup et al. 2016).

DOI: 10.1201/9781003508724-10

FIGURE 10.1 Photograph of *Curcuma comosa* Roxb.

TABLE 10.1
Phytochemicals Isolated from *C. comosa*

Name	Structure	Examples of Potential Uses in Medicine
(3R)-1,7-diphenyl-(4E,6E)-4,6-heptadien-3-ol, or compound 049, or ASPP 049	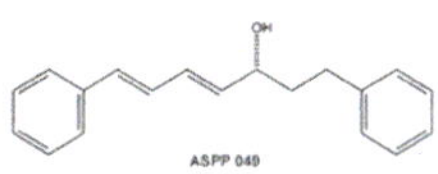 ASPP 049	Anti-inflammatory activity resulting in neuroprotective activity (Thampithak et al. 2009) Hypolipidaemic activity and improved insulin sensitivity in estrogen-deprived state, i.e., improvement in glucose and lipid metabolism (Prasannarong et al. 2012)
(3S)-7-(3,4-dihydroxyphenyl)-1-phenyl-(1E)-1-hepten-3-ol, or ASPP 092	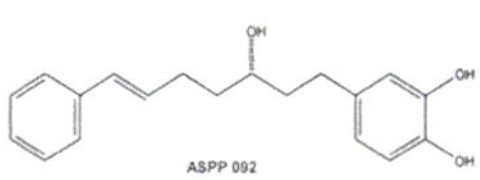 ASPP 092	Anti-oxidant activity resulting in anti-apoptotic and neuroprotective activity (Vattanarongkup et al. 2016)
(4E,6E)-1,7-diphenylhepta-4,6-dien-3-one	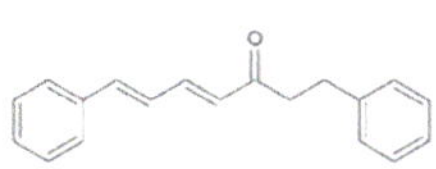 (4*E*,6*E*)-1,7-diphenylhepta-4,6-dien-3-one	Spatial learning and memory (Su et al. 2010)

10.3 MEDICINAL PROPERTIES OF PLANT EXTRACTS AND CHEMICAL CONSTITUENTS

10.3.1 Neuroprotective Role

Neurodegenerative disorders, such as Parkinson's and Alzheimer's diseases, are believed to involve chronic activation of microglia, resulting in overproduction of pro-inflammatory mediators. Thampithak et al. used a lipopolysaccharide (LPS)-treated, highly aggressively proliferating immortalised (HAPI) rat microglial cell

0.1, 1, and 10 μM, respectively, relative to the control (10.1 ± 0.9 mg/dL) ($P < 0.01$). Furthermore, they utilised gene expression profiling to show that the expression of numerous genes in the Wnt/β-catenin signalling pathway, which is a negative regulator of adipocyte differentiation in hBMSCs, was increased by DPHD treatment, which also elevated the levels of extracellular signal-regulated kinases 1 and 2 (ERK1/2) and glycogen synthase kinase-3 beta (GSK-3β) that connect the oestrogen receptor (ER) and Wnt/β-catenin signalling pathways. Since bone marrow adiposity is one of the crucial factors that aggravate osteoporosis in post-menopause, this discovery indicates the potential role of DPHD in its prevention.

10.3.3 Spatial Memory

Since the extract and some purified compounds from *C. comosa* reportedly have oestrogen-like effects and oestrogen improves learning in some animals, Su et al. determined if *C. comosa* and oestrogen have similar beneficial effects on spatial learning and memory (2010). They used ovariectomised Wistar rats and orally administered *C. comosa* hexane extract containing 0.165 mg of (4E,6E)-1,7-diphenylhepta-4,6-dien-3-one per mg of the crude extract, at doses of 250 or 500 mg/kg body weight, with 17β-estradiol (10 μg/kg body weight, subcutaneously) used as a positive control. A Morris water maze was used to test for spatial learning and memory thirty days after the initiation of treatment, with re-testing every 30 days and a final probe trial on day 119. Whereas significant memory impairment for locating the platform in the water maze was exhibited by OVX rats from day 67 after the surgery onwards, *C. comosa* or oestrogen treatment offered significant protection from this decline in cognitive function, which for *C. comosa* was more extensive at the higher dose. Therefore, long-term treatment with *C. comosa* has favourable effects on learning and memory function in rats and, by extension, potentially in postmenopausal women.

In a second study (2011), Su et al. examined the effects of *C. comosa* hexane extract on spatial memory and oestrogen receptors (ER) via ER mRNA expression in the hippocampus of Wistar rats that were trained for five days in a Morris water maze before ovariectomy. *C. comosa* hexane extract was then intraperitoneally administered at 250 and 500 mg/kg body weight daily for four weeks, with 10 μg/kg oestradiol used as the positive control. The animals' latency to the platform was increased significantly by OVX, and the memory impairments were reversed by oestradiol, and both doses of *C. comosa* extract. Reverse transcription polymerase chain reaction demonstrated a selective increase of the ER alpha but not beta subtype in the hippocampus for *C. comosa* treatment, with an increase in both for oestradiol treatment. *C. comosa* did not have the above effects in the sham OVX rats. They inferred that the influence of *C. comosa* hexane extract on the memory changes in the OVX rats might be related selectively to ER alpha in the hippocampus.

10.4 CONCLUSIONS

C. comosa has been demonstrated to exhibit anti-inflammatory activity resulting in neuroprotective activity, antioxidant activity resulting in anti-apoptotic and neuroprotective activity, hypolipidemic activity, and improved insulin sensitivity in oestrogen-deprived states, along with favourable effects on spatial learning and memory, among other activities, based on the scientific literature. These findings justify the traditional use of *C. comosa* in Thai folk medicine and offer promise for further drug development.

REFERENCES

Intapad S, Saengsirisuwan V, Prasannarong M, et al., Long-term effect of phytoestrogens from *Curcuma comosa* Roxb. on vascular relaxation in ovariectomized rats. 2012. *J Agric Food Chem*, 60(3): 758–64.

Khin Aung ZM, Jantaratnotai N, Piyachaturawat P, et al., A pure compound from *Curcuma comosa* Roxb. protects neurons against hydrogen peroxide-induced neurotoxicity via the activation of Nrf-2. 2022. *Heliyon*, 8(11): e11228.

Piyachaturawat P, Charoenpiboonsin J, Toskulkao C, et al., Reduction of plasma cholesterol by Curcuma comosa extract in hypercholesterolaemic hamsters. 1999. *J Ethnopharmacol*, 66(2): 199–204.

Piyachaturawat P, Ercharuporn S, Suksamrarn A. Estrogenic activity of Curcuma comosa extract in rats. 1995. *Asian Pac J Pharmacol*, 10: 121–126.

Prasannarong M, Saengsirisuwan V, Piyachaturawat P, et al., Improvements of insulin resistance in ovariectomized rats by a novel phytoestrogen from *Curcuma comosa* Roxb. 2012. *BMC Complement Altern Med*, 12: 28.

Su J, Sripanidkulchai B, Sripanidkulchai K, et al., Effect of Curcuma comosa and estradiol on the spatial memory and hippocampal estrogen receptor in the post-training ovariectomized rats. 2011. *J Nat Med*, 65(1): 57–62.

Su J, Sripanidkulchai K, Wyss JM, et al., Curcuma comosa improves learning and memory function on ovariectomized rats in a long-term Morris water maze test. 2010. *J Ethnopharmacol*, 130(1): 70–5.

Suksamrarn A, Ponglikitmongkol M, Wongkrajang K, et al., Diarylheptanoids, new phytoestrogens from the rhizomes of Curcuma comosa: Isolation, chemical modification and estrogenic activity evaluation. 2008. *Bioorg Med Chem Lett*, 16: 6891–6902.

Sutjarit N, Sueajai J, Boonmuen N, et al., Curcuma comosa reduces visceral adipose tissue and improves dyslipidemia in ovariectomized rats. 2018. *J Ethnopharmacol*, 215: 167–175.

Sutjarit N, Thongon N, Weerachayaphorn J, et al., Inhibition of adipogenic differentiation of human bone marrow-derived mesenchymal stem cells by a phytoestrogen diarylheptanoid from Curcuma comosa. 2020. *J Agric Food Chem*, 68(37): 9993–10002.

Thampithak A, Jaisin Y, Meesarapee B, et al., Transcriptional regulation of iNOS and COX-2 by a novel compound from Curcuma comosa in lipopolysaccharide-induced microglial activation. 2009. *Neurosci Lett*, 462(2): 171–5.

Vattanarongkup J, Piyachaturawat P, Tuchinda P, et al., Protective effects of a diarylheptanoid from Curcuma comosa against hydrogen peroxide-induced astroglial cell death. 2016. *Planta Med*, 82(17): 1456–1462.

Xu F, Nakamura S, Qu Y, et al., Structures of new sesquiterpenes from Curcuma comosa. 2008. *Chem Pharm Bull (Tokyo)*, 56: 1710–16.

11 Traditional Thai Medicinal Properties of *Cleistocalyx nervosum* var. paniala

Rajasekharan Sharika, Mani Iyer Prasanth, Bhagavathi Sundaram Sivamaruthi, and Tewin Tencomnao

11.1 INTRODUCTION

Medicinal plant-based approaches to treat diseases have been followed in many countries since time immemorial. Even with modern medicinal facilities, plant-based approaches have been utilised and proposed to effectively treat many diseases by identifying and formulating the active biomolecule(s) present in fruits, stems, leaves, flowers, and seeds (Kumar et al. 2019). The traditional Thai treatment mainly relies on the phytochemical bounty of the forest flora within various plants indigenously available in Thailand and the Asian continent. Among the indigenously available plants, the fruit of *Cleistocalyx nervosum* var. paniala (Figure 11.1) has been traditionally used to prepare jams, fruit drinks, wine, and nectar by local Thais (Manosroi et al. 2015).

C. nervosum belongs to the Myrtaceae family of Plant Kingdom, widely found along the northern region of Thailand, and is commonly known as Ma-Kiang. It produces an edible, slightly sour-tasting fruit (flesh and seeds) that is a source of high amounts of polyphenols and flavonoids (Chaisawadi and Methawiriyasilp 2007; Nuengchamnong and Ingkaninan 2009; Sriwanthana et al. 2007). Owing to its high phytochemical content such as flavonoids, phenolics, and proanthocyanidins, the fruit extracts have been marketed in the form of functional health food, cosmetic ingredients, and health drinks (Nuengchamnong and Ingkaninan 2009; Poontawee et al. 2016).

DOI: 10.1201/9781003508724-11

FIGURE 11.1 Photograph of *Cleistocalyx nervosum* var. paniala. (A) Flowers, (B) berries (green and unripe), and (C) Berries (dark red and ripe).

11.2 MEDICINAL PROPERTIES OF PLANT EXTRACT AND CHEMICAL CONSTITUENTS

The notion "prevention is better than cure" has gained impetus with increased incidences of various degenerative diseases, making people critically analyse the nutritional value of their food choices. The advent of a scientific approach to disease mechanisms has enlightened the importance of phytochemicals such as polyphenolics and flavonoids in the diet as a source of potent biomolecules that have a variety of roles such as anti-inflammatory, anti-cancerous, anti-heavy metal toxicity, anti-microbial activity, anti-ageing, neuroprotectivity, and immunomodulation (Patthamakanokporn et al. 2008; Prasanth et al. 2019). The fruit, seed, and leaf extracts of *C. nervosum* have been reported to have high amounts of phytochemical contents of therapeutic value.

HPLC analysis coupled with ESI-MS/MS-DPPH assay identification revealed the presence of antioxidants such as gallic acid, ellagic acid, methoxymethylgallate, and caffeoylquinic acid present in *C. nervosum* fruit wine (Nuengchamnong and Ingkaninan 2009). The presence of anthocyanin cyanidin 3-glucoside is attributed to the rich reddish-purple colour of the ripened fruit (Jansom et al. 2008; Chaiyasut et al. 2016).

11.2.1 Antioxidant Property

Reactive oxygen species (ROS) include both free radical and non-free radical molecules that can generate activated oxygen species (Manosroi et al. 2015; Patthamakanokporn et al. 2008) and behave as a double-edged sword, as it plays a crucial role in the innate immune response. However, an overall increase in free radicals can also cause various chronic diseases. ROS can oxidatively modify macromolecular structures, adversely affecting human health (Schieber and Chandel 2014).

C. nervosum fruit extract has been reported to show the highest oxygen radical absorbance capacity (ORAC), making it an excellent source of antioxidants compared to other widely consumed fruits in Thailand (Patthamakanokporn et al. 2008). *In vivo,* analysis in two separate studies revealed the practical impact of antioxidants in the aqueous and methanolic extracts of *C. nervosum* fruit and seed on rat liver cells (Taya et al. 2009). A study by Chaiyasut et al. 2016 revealed that the fruit extract of *C. nervosum* had the highest amount of anthocyanin, cyanidin 3-glucoside, which showed increased antioxidant activity compared to other locally consumed fruits. Our recent study also reports the activation of antioxidant pathways in wild-type *C. elegans*, mediated by skn-1 and sir-2.1, during treatment with *C. nervosum* (Prasanth et al. 2019).

11.2.2 Anti-ageing Activity

Chronological ageing is a natural phenomenon that reduces overall physiological and biological function. In contrast, photoageing is caused by the exposure of the skin to UV radiation from the sun, leading to the rapid generation and accumulation of reactive oxygen species (Rittié and Fisher 2015). Studies have revealed the importance of plant-derived phytochemicals such as phenolic compounds, ascorbic acid, and carotenoids in protecting the skin by influencing several signaling pathways (Petruk et al. 2018). In this aspect, *C. nervosum* is rich in many polyphenols and flavonoids such as gallic acid, ellagic acid, methoxymethylgallate, caffeoylquinic acid, and cyanidin 3-glucoside. Gallic acid (a naturally occurring low molecular weight triphenolic compound) and cyanidin 3-glucoside have been reported to prevent UV-B- mediated photoageing effects via downregulation of melanogenic genes, upregulation of TGF-β-induced procollagen production, and scavenging of UV-B-induced free radicals, respectively (Ding et al. 2006; Kumar et al. 2013; Hwang et al. 2014; Kahkeshani et al. 2019). Recently, Asnaashari et al. 2019 reported the effectiveness of gallic acid and its ester form methyl methoxy gallate as an alternative to synthetic antioxidants (Asnaashari et al. 2019).

11.2.3 Anticarcinogenic Activity

Extensive *in vivo* studies reveal gallic acid and its derivatives as potent molecules for antitumour activity, attributed to their antioxidant and prooxidant effects. The inhibition of invasion, proliferation, and metastasis of cancer by gallic acid is achieved via slowdown of cell division by regulating cell cycle–related proteins such as Cyclins and the induction of p27KIP enzyme, as well as inhibition of CDK and matrix metalloproteinase activity (Giftson et al. 2010; Subramanian et al. 2014; Balch et al. 2009; Liao et al. 2012; Nemec et al. 2016; Nemec et al. 2016; Huang et al. 2012). Jagan et al. 2008 reported a decrease in tumour size as well as serum levels of markers such as aspartate transaminase (AST), alanine transaminase (ALT), lactate dehydrogenase

(LDH), alkaline phosphatase (ALP), and gamma-glutamyl transferase (GGT) during hepatocellular carcinoma.

Flavonoids like anthocyanins are also known to demonstrate protective effects against viruses, bacteria, and fungi. Several *in vitro, in vivo,* and human epidemiological studies have revealed chemopreventive and chemo-protective properties against multiple types of cancer (Diaconeasa et al. 2018) by triggering selective cell cycle arrest, thereby preventing cancer cell proliferation, as well as an internal mitochondrial pathway by modulating caspase-dependent anti- and pro-apoptotic proteins, and external death receptor pathway via modulating FAS and FASL, resulting in cellular apoptosis. It is also known to suppress angiogenesis through inhibiting H_2O_2 and TNF-α (tumour necrosis factor)-induced VGF expression. Similar to gallic acid, anthocyanins also inhibit metastasis by reducing the expression of matrix metalloproteins (MMPs) and urokinase plasminogen activators (Rugină et al. 2012; Diaconeasa et al. 2015; Malik et al. 2003; Wang and Stoner 2008; Reddivari et al. 2007; Chang et al. 2005; Hientz et al. 2017; Bagchi et al. 2004; Lamy et al. 2007; Brandstetter et al. 2001; Diaconeasa et al. 2018). Anthocyanin, cyanidin and 3-glucoside was reported to be effective against *in vitro* and *in vivo* HER2 type (human epidermal growth factor receptor 2) breast cancer cells through tumour suppression and inhibition of angiogenesis (Diaconeasa et al. 2018).

11.2.4 Other Protective and Antagonistic Activity of *C. nervosum* Extracts

The dietary inclusion of plant-derived flavonoids and non-flavonoid phenolic phytochemicals via fruits and vegetables provides several benefits to human health. Several *in vitro* and *in vivo* studies have shown the importance of these components in lipid and glucose metabolism (Kang et al. 2016). Ellagic acid, one of the abundant phytochemicals identified in *C. nervosum* extracts, is a bioactive component reported to provide a synergistic effect with other dietary components to aid metabolism. Though ellagic acid has a low absorption rate in the gut, its metabolic products, urolithins (metabolised by the gut microbiota), are effectively absorbed and bioavailable due to their high lipophilicity (Espín et al. 2007; García-Villalba et al. 2013; González-Sarrías et al. 2015; Mertens-Talcott et al. 2006). The absorbed ellagic acids are mostly converted to methyl esters, dimethyl esters, and glucuronides. Hence, ellagic acid–mediated human health benefits are, to an extent, attributed to the urolithins. However, extensive studies are needed. Urolithins are formed by the breaking of the lactone ring followed by the removal of a carboxy group to form urolithin D and urolithin C, iso-urolithin A, and urolithin B by losing 1, 2, and 3 hydroxyl groups, respectively (Seeram et al. 2006; González-Sarrías et al. 2015; Mertens-Talcott et al. 2006). Ellagic acid and its metabolites (urolithins) are reported to be beneficial in attenuating obesity and associated metabolic dysfunction via epigenetic and immune modulation (Kang et al. 2016).

Plant extracts have been reported to provide neuroprotection against chronic and acute neurodegenerative diseases such as Alzheimer's, Parkinson's, Huntington's,

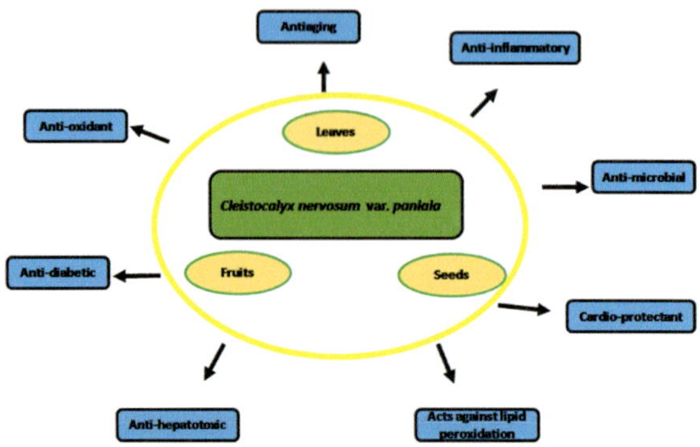

FIGURE 11.2 Medicinal properties of *Cleistocalyx nervosum* var. paniala.

stroke, as well as amyotrophic lateral sclerosis. The underlying pathophysiology of these degenerative diseases of neurons is mainly due to protein aggregation, inflammation, excitotoxicity, oxidative stress, and neurotoxicity, which are interlinked. Traditionally used plant extracts of *Ginkgo biloba, Panax ginseng, Curcuma longa, Bacopa monnieri,* and *Salvia officinalis* have been reported to have neuroprotective effects by several studies due to the presence of polyphenols, terpenoids, and alkaloids (Iriti et al. 2010; Adewusi et al. 2010; Nieoullon 2011; Reeve et al. 2014; Hindle 2010; Rates 2001; Harvey 2008; Ekor 2014; Chen and Pan 2014; Kumar and Khanum 2012; Elufioye et al. 2017; Kahkeshani et al. 2019). *C. nervosum* has also been reported to show neuroprotective effects in both *in vitro* (against glutamate-induced toxicity in HT22 cell lines) and *in vivo* (significant increase in the lifespan of transgenic *C. elegans* model for Alzheimer's) studies (Sukprasansap et al. 2017; Prasanth et al. 2019).

11.3 CONCLUSIONS

The scientific literature reveals *C. nervosum* as an excellent source of potent bioactive components against metabolic and degenerative disorders. Extensive *in vivo* studies will clarify these bioactive compounds' full potential and mechanism (Figure 11.2).

REFERENCES

Adewusi EA, Moodley N, Steenkamp V (2010) Medicinal plants with cholinesterase inhibitory activity: A review. *Afr J Biotechnol.* 9(49): 8257–8276.

Asnaashari M, Farhoosh R, Farahmandfar R (2019) Preservation of gallic acid and methyl gallate on purified Kilka fish oil oxidation by Rancimat. *Food Sci Nutr.* 7(12): 4007–4013.

Bagchi D, Sen CK, Bagchi M, Atalay M (2004) Anti-angiogenic, antioxidant, and anticarcinogenic properties of a novel anthocyanin-rich berry extract formula. *Biochemistry (Moscow).* 69(1): 75–80.

Balch CM, Gershenwald JE, Soong SJ, Thompson JF, Atkins MB, et al. (2009) Final version of 2009 AJCC melanoma staging and classification. *J Clin Oncol.* 27(36): 6199.

Brandstetter H, Grams F, Glitz D, Lang A, Huber R, et al. (2001) The 1.8-Å crystal structure of a matrix metalloproteinase 8-barbiturate inhibitor complex reveals a previously unobserved mechanism for collagenase substrate recognition. *J Biol Chem.* 276(20): 17405–17412.

Chaisawadi S, Methawiriyasilp W (2007) Clean production of commercial 'Makiang' juice processing for medicinal herbs and health benefits. *International Workshop on Medicinal and Aromatic Plants* 786: 201–208.

Chaiyasut C, Sivamaruthi BS, Pengkumsri N, Sirilun S, Peerajan S, et al. (2016) Anthocyanin profile and its antioxidant activity of widely used fruits, vegetables, and flowers in Thailand. *Asian J Pharm Clin Res.* 9(6): 218–224.

Chang YC, Huang HP, Hsu JD, Yang SF, Wang CJ (2005) Hibiscus anthocyanins rich extract-induced apoptotic cell death in human promyelocytic leukemia cells. *Toxicol Appl Pharmacol.* 205(3): 201–212.

Chen X, Pan W (2014) The treatment strategies for neurodegenerative diseases by integrative medicine. *Integr Med Int.* 1(4): 223–225.

Diaconeasa Z, Leopold L, Rugină D, Ayvaz H, Socaciu C (2015) Antiproliferative and antioxidant properties of anthocyanin rich extracts from blueberry and blackcurrant juice. *Int J Mol.* 16(2): 2352–2365.

Diaconeasa ZM, Frond AD, Ştirbu I, Rugina D, Socaciu C (2018) Anthocyanins-smart molecules for cancer prevention. *Phytochemicals: Source of Antioxidants and Role in Disease Prevention*, 75–94.

Ding M, Feng R, Wang SY, Bowman L, Lu Y, et al. (2006) Cyanidin-3-glucoside, a natural product derived from blackberry, exhibits chemopreventive and chemotherapeutic activity. *J Biol Chem.* 281(25): 17359–17368.

Ekor M (2014) The growing use of herbal medicines: Issues relating to adverse reactions and challenges in monitoring safety. *Front Pharmacol.* 4: 177.

Elufioye TO, Berida TI, Habtemariam S (2017) Plants-derived neuroprotective agents: Cutting the cycle of cell death through multiple mechanisms. *Evid Based Complement Alternat Med.* 2017: 3574012.

Espín JC, González-Barrio R, Cerdá B, López-Bote C, Rey AI, Tomás-Barberán FA (2007) Iberian pig as a model to clarify obscure points in the bioavailability and metabolism of ellagitannins in humans. *J Agric Food Chem.* 55(25): 10476–10485.

García-Villalba R, Beltrán D, Espín JC, Selma MV, Tomás-Barberán FA (2013) Time course production of urolithins from ellagic acid by human gut microbiota. *J Agric Food Chem.* 61(37): 8797–8806.

Giftson JS, Jayanthi S, Nalini N (2010) Chemopreventive efficacy of gallic acid, an antioxidant and anticarcinogenic polyphenol, against 1, 2-dimethyl hydrazine induced rat colon carcinogenesis. *Invest New Drug.* 28(3): 251–259.

González-Sarrías A, García-Villalba R, Núñez-Sánchez MÁ, Tomé-Carneiro J, Zafrilla P, et al. (2015) Identifying the limits for ellagic acid bioavailability: A crossover pharmacokinetic study in healthy volunteers after consumption of pomegranate extracts. *J Funct Foods* 19: 225–235.

Harvey AL (2008) Natural products in drug discovery. *Drug Discov Today.* 13(19–20): 894–901.

Hientz K, Mohr A, Bhakta-Guha D, Efferth T (2017) The role of p53 in cancer drug resistance and targeted chemotherapy. *Oncotarget* 8(5): 8921.

Hindle JV (2010) Ageing, neurodegeneration and Parkinson's disease. *Age Ageing.* 39(2): 156–161.

Huang PJ, Hseu YC, Lee MS, Kumar KS, Wu CR, et al. (2012). In vitro and in vivo activity of gallic acid and Toona sinensis leaf extracts against HL-60 human premyelocytic leukemia. *Food Chem Toxicol.* 50(10): 3489–3497.

Hwang E, Park SY, Lee HJ, Lee TY, Sun ZW, Yi TH (2014) Gallic acid regulates skin photoaging in UVB exposed fibroblast and hairless mice. *Phytother Res.* 28(12): 1778–1788.

Iriti M, Vitalini S, Fico G, Faoro F (2010) Neuroprotective herbs and foods from different traditional medicines and diets. *Molecules.* 15(5): 3517–3555.

Jagan S, Ramakrishnan G, Anandakumar P, Kamaraj S, Devaki T (2008) Antiproliferative potential of gallic acid against diethylnitrosamine-induced rat hepatocellular carcinoma. *Mol Cell Biochem.* 319(1–2): 51.

Jansom C, Bhamarapravati S, Itharat A (2008) Major anthocyanin from ripe berries of Cleistocalyx nervosum var. paniala. *Thammasat Medical J.* 8(3): 365.

Kahkeshani N, Farzaei F, Fotouhi M, Alavi SS, Bahramsoltani R, et al. (2019). Pharmacological effects of gallic acid in health and diseases: A mechanistic review. *Iran J Basic Med Sci.* 22(3): 225.

Kang I, Buckner T, Shay NF, Gu L, Chung S (2016) Improvements in metabolic health with consumption of ellagic acid and subsequent conversion into urolithins: Evidence and mechanisms. *Adv Nutr.* 7(5): 961–972.

Kumar GP, Khanum F (2012) Neuroprotective potential of phytochemicals. *Pharmacogn Rev.* 6(12): 81.

Kumar KS, Vani MG, Wang SY, Liao JW, Hsu LS, et al. (2013) In vitro and in vivo studies disclosed the depigmenting effects of gallic acid: A novel skin lightening agent for hyperpigmentary skin diseases. *Biofactors* 39(3): 259–270.

Kumar SJ, Tungphatthong C, Sukrong S (2019) Mitigating the impact of admixtures in Thai herbal products. *Front Pharmacol.* 10: 1205.

Lamy S, Lafleur R, Bédard V, Moghrabi A, Barrette S, Gingras D, Béliveau R (2007) Anthocyanidins inhibit migration of glioblastoma cells: Structure-activity relationship and involvement of the plasminolytic system. *J Cell Biochem.* 100(1): 100–111.

Liao CL, Lai KC, Huang AC, Yang JS, Lin JJ, et al. (2012). Gallic acid inhibits migration and invasion in human osteosarcoma U-2 OS cells through suppressing the matrix metalloproteinase-2/-9, protein kinase B (PKB) and PKC signaling pathways. *Food Chem Toxicol.* 50(5): 1734–1740.

Malik M, Zhao C, Schoene N, Guisti MM, Moyer MP, Magnuson BA (2003) Anthocyanin-rich extract from Aronia meloncarpa E. induces a cell cycle block in colon cancer but not normal colonic cells. *Nutr Cancer.* 46(2): 186–196.

Manosroi J, Chankhampan C, Kumguan K, Manosroi W, Manosroi A (2015) In vitro antiaging activities of extracts from leaves of Ma Kiang (Cleistocalyx nervosum var. paniala). *Pharm Biol.* 53(6): 862–869.

Mertens-Talcott SU, Jilma-Stohlawetz P, Rios J, Hingorani L, Derendorf H (2006) Absorption, metabolism, and antioxidant effects of pomegranate (Punica granatum L.) polyphenols after ingestion of a standardised extract in healthy human volunteers. *J Agric Food Chem.* 54(23): 8956–8961.

Nemec MJ, Kim H, Marciante AB, Barnes RC, Talcott ST, Mertens-Talcott SU (2016) Pyrogallol, an absorbable microbial gallotannins-metabolite and mango polyphenols (Mangifera Indica L.) suppress breast cancer ductal carcinoma in situ proliferation in vitro. *Food Funct.* 7(9): 3825–3833.

Nieoullon A (2011) Neurodegenerative diseases and neuroprotection: Current views and prospects. *J Appl Biomed.* 9(4): 173–183.

Nuengchamnong N, Ingkaninan K (2009) On-line characterisation of phenolic antioxidants in fruit wines from family Myrtaceae by liquid chromatography combined with electrospray ionisation tandem mass spectrometry and radical scavenging detection. *LWT-Food Sci Tech.* 42(1): 297–302.

Patthamakanokporn O, Puwastien P, Nitithamyong A, Sirichakwal PP (2008) Changes of antioxidant activity and total phenolic compounds during storage of selected fruits. *J Food Compos Anal.* 21(3): 241–248.

Petruk G, Del Giudice R, Rigano MM, Monti DM (2018) Antioxidants from plants protect against skin photoaging. *Oxid Med Cell Longev.* 2018: 1454936.

Poontawee W, Natakankitkul S, Wongmekiat O (2016) Protective effect of Cleistocalyx nervosum var. paniala fruit extract against oxidative renal damage caused by cadmium. *Molecules* 21(2): 133.

Prasanth MI, Brimson JM, Chuchawankul S, Sukprasansap M, Tencomnao T (2019) Antiaging, stress resistance, and neuroprotective efficacies of Cleistocalyx nervosum var. paniala fruit extracts using Caenorhabditis elegans model. *Oxid Med Cell Longev.* 2019: 7024785.

Rates SMK (2001) Plants as source of drugs. *Toxicon.* 39(5): 603–613.

Reddivari L, Vanamala J, Chintharlapalli S, Safe SH, Miller Jr JC (2007) Anthocyanin fraction from potato extracts is cytotoxic to prostate cancer cells through activation of caspase-dependent and caspase-independent pathways. *Carcinogenesis* 28(10): 2227–2235.

Reeve A, Simcox E, Turnbull D (2014) Ageing and Parkinson's disease: Why is advancing age the biggest risk factor?. *Ageing Res Rev.* 14: 19–30.

Rittié L, Fisher GJ (2015) Natural and sun-induced aging of human skin. *Cold Spring Harb Perspect Med.* 5(1): a015370.

Rugină D, Sconţa Z, Leopold L, Pintea A, Bunea A, Socaciu C (2012) Antioxidant activities of chokeberry extracts and the cytotoxic action of their anthocyanin fraction on HeLa human cervical tumor cells. *J Med Food.* 15(8): 700–706.

Schieber M, Chandel NS (2014) ROS function in redox signaling and oxidative stress. *Curr Biol.* 24(10): R453–R462.

Seeram NP, Henning SM, Zhang Y, Suchard M, Li Z, Heber D (2006) Pomegranate juice ellagitannin metabolites are present in human plasma and some persist in urine for up to 48 hours. *J Nutr.* 136(10): 2481–2485.

Sriwanthana B, Treesangsri W, Boriboontrakul B, Niumsakul S, Chavalittumrong P (2007) In vitro effects of Thai medicinal plants on human lymphocyte activity. *Songklanakarin J Sci Technol.* 29: 1.

Subramanian V, Venkatesan B, Tumala A, Vellaichamy E (2014) Topical application of Gallic acid suppresses the 7, 12-DMBA/Croton oil induced two-step skin carcinogenesis by modulating antioxidants and MMP-2/MMP-9 in Swiss albino mice. *Food Chem Toxicol.* 66: 44–55.

Sukprasansap M, Chanvorachote P, Tencomnao T (2017) Cleistocalyx nervosum var. paniala berry fruit protects neurotoxicity against endoplasmic reticulum stress-induced apoptosis. *Food Chem Toxicol.* 103: 279–288.

Taya S, Punvittayagul C, Chewonarin T, Wongpoomchai R (2009) Effect of aqueous extract from Cleistocalyx nervosum on oxidative status in rat liver. *Thai J Toxicol.* 24(2): 101–115.

Wang LS, Stoner GD (2008) Anthocyanins and their role in cancer prevention. *Cancer Lett.* 269(2): 281–229.

12 Kratom

A Beneficial Traditional Thai Medicine or a Harmful Narcotic Drug?

Clerance Su Yee Cheong and Tewin Tencomnao

12.1 INTRODUCTION

The *Mitragyna* genus, belonging to the Rubiaceae family, comprises ten species, six of which are found in Asia and four in Africa. The most prevalent species in Thailand are *Mitragyna speciosa*, *Mitragyna diversifolia*, *Mitragyna hirsuta*, *Mitragyna rotundifolia*, and *Mitragyna parvifolia* (Sengnon et al., 2023). *Mitragyna speciosa* (Korth.) Havil. (*M. speciosa*) is extensively studied among all these species due to its narcotic properties as an opium substitute (Saidin et al., 2015). *M. speciosa*, known as kratom, is a tropical medicinal plant native to peninsular Thailand, Malaysia, Indonesia, and other Southeast Asian countries. Among the Thai Indigenous community, it is also known as "*kakuam, kraton, ithang* or *thom*" (Yusoff et al., 2016).

Kratom boasts a long history of ethnomedicinal use in Thailand, which is evident in Thai Materia Medica. For example, prasakratom, mainly composed of kratom leaves, is used as a traditional regimen to treat diarrhoea (Tsuchiya et al., 2002; Sengnon et al., 2023). Other than diarrhoea, kratom has also been used in folk medicine to relieve pain and fatigue (Singh et al., 2022; Mat et al., 2023). To this day, native populations, specifically in rural areas, either consume the fresh leaves directly or prepare kratom leaves into extract before consumption (Singh, Müller and Vicknasingam, 2014; Singh, Abdullah, et al., 2018).

Kratom is commonly used for its stimulating and analgesic properties, with varying effects depending on dosage, which may also include psychostimulant and opiate-like effects. The plant concoction is often utilised as a cheaper alternative to heroin for treating opiate withdrawal (Singh et al., 2022). Its easy accessibility has led to widespread and unrestrained usage despite limited awareness of its potential adverse health effects. Anorexia, weight loss, jerky limb movements, anger, aching muscles and bones, sleeplessness, and psychosis have all been described as withdrawal symptoms (Kumarnsit, Keawpradub and Nuankaew, 2007; Cheaha et al., 2015). However, despite these negative impacts, moderate kratom consumption may

DOI: 10.1201/9781003508724-12

confer benefits such as anti-inflammatory, antidiabetic, neuroprotection, and gastrointestinal protection (Hossain et al., 2023).

12.2 BOTANICAL DISTRIBUTION AND DESCRIPTION

Under medium to full sunlight and humid weather conditions, *M. speciosa* thrives as an evergreen tree, capable of reaching heights of up to 30 metres. Thai kratom leaves are typically categorised as either red- or green-veined, with some featuring a serrated apex (Ngernsaengsaruay et al., 2022). The prevalence of active pharmacological compounds in leaves with crimson, green, and apical veins and serrated edges remains unclear (Sengnon et al., 2023). Nevertheless, indigenous people in southern Thailand believe green leaves are energising, while crimson leaves possess medicinal properties (Saingam et al., 2013). However, current molecular techniques have not yet succeeded in distinguishing the genetic makeup between different leaf types. Figure 12.1 illustrates the morphology of the kratom plant and its leaves.

Kratom is naturally distributed across Thailand, with most of its abundance occurring in the southern regions near riverbanks and in national forests (Saingam et al., 2013). Nam Phu subdistrict, Ban Na San, and Surat Thani are regions authorised by the Narcotic Crop Survey and Monitoring Institute (NCS), the Office of the Narcotics Control Board (ONCB) of Thailand, for the cultivation of kratom trees (Sengnon et al., 2023).

12.3 ETHNOMEDICAL USES OF KRATOM

The southern region of Thailand is renowned as a hotspot for *M. speciosa*, with strong historical and cultural traditions among the natives in utilising this plant. Additionally, the locals have also used this plant as an ethnomedicine. Farmers

FIGURE 12.1 Young kratom plant, image kindly provided by Assoc.Prof.Dr Theerakamol Pengsakul, Prince of Songkla University, Thailand.

commonly use the leaves of *M. speciosa* as a stimulant to alleviate fatigue and muscular pain resulting from prolonged work in the field (Suwanlert, 1975). While the leaves are primarily utilised for their ethnomedicinal purposes, the bark, stem, and root of *M. speciosa* are not commonly employed for medicinal use, suggesting that the majority of known ethnomedicinal properties are attributed to the leaves, possibly due to the concentration of beneficial bioactive compounds therein.

Beyond its role as an analgesic agent, kratom offers benefits for inflammation, cancer, neurodegenerative disorders, gastrointestinal issues, and relieving drug withdrawal symptoms (Singh, Müller, and Vicknasingam, 2014; Yusoff et al., 2016; Arneth, 2018; H. Q. Li et al., 2019).

12.4 PHYTOCHEMISTRY OF KRATOM

The therapeutic efficacy of medicinal plants is commonly attributed to their phytochemical properties. Kratom has garnered research interest due to the presence of pharmacologically active compounds beneficial for various clinical disorders (Firmansyah, Sundalian and Taufiq, 2021). The phytochemical constituents of kratom reportedly include 86 secondary metabolites, such as alkaloids, flavonoids, terpenoids, polyphenols, triterpenoids, and saponins (Phongprueksapattana et al., 2008; Veeramohan et al., 2023).

12.4.1 Alkaloids Found in Kratom

The therapeutic efficacy of medicinal plants is commonly attributed to their phytochemical properties. Kratom has garnered research interest due to the presence of pharmacologically active compounds beneficial for various clinical disorders (Firmansyah, Sundalian and Taufiq, 2021). The phytochemical constituents of kratom reportedly include 86 secondary metabolites, such as alkaloids, flavonoids, terpenoids, polyphenols, triterpenoids, and saponins (Phongprueksapattana et al., 2008; Veeramohan et al., 2023) (Figure 12.2).

There are numerous probable and confirmed isolated compounds from kratom which exert pharmacological effects: (+)-N-(methoxycarbonyl)-N-norboldine (antibacterial) (Qing et al., 2017), 6-Deoxyfagomine, acronycine (anticancer) (Rocca et al., 2015), adouetine X (analgesic) (Trevisan et al., 2009), ajmalicine (antihypertensive, anticancer, sedative, anticonvulsant) (Kashyap et al., 2020; Sun, Ma and Zhan, 2023), ajmaline (antiarrhythmic) (Postema et al., 2010), akuammidine (analgesic, stimulant) (Creed et al., 2021), brucine (analgesic, anticancer, anti-inflammatory) (Yin et al., 2003; Agrawal et al., 2011), ciliaphylline (anti-diarrheal, antibacterial, antitussive), corynantheidine (analgesic, antihypertensive) (Váradi et al., 2016), 9-hydroxycorynantheidine (partial opioid receptor agonist) (Matsumoto, Takayama, et al., 2006), corynoxeine (neuroprotective, anti-locomotive, antidepressant) (Qi et al., 2015; Sun et al., 2024), daphniphylline (alkaloids), dauricine (anti-inflammatory, anti-arrhythmic, antitumour, neuroprotective) (Zhang, Ren, and Qiu, 2018; Wang et al., 2020), epicatechin (anti-inflammatory, antioxidant, antibacterial, cardiac-protective, antitumour) (García-Díez et al., 2024; Osorio-Cruz et al., 2024), eseramine

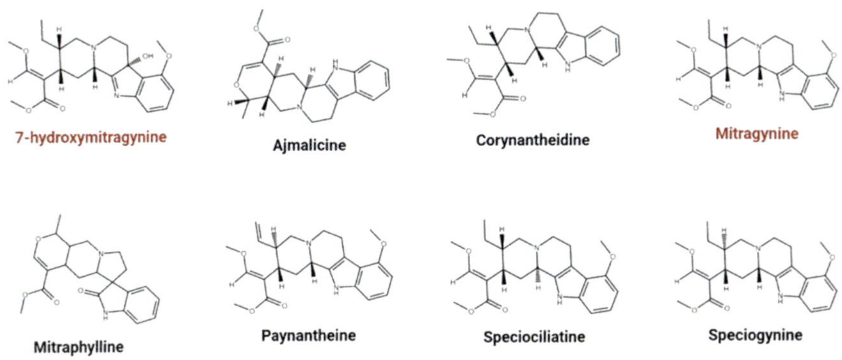

FIGURE 12.2 Phytochemical structure of the most prominent compounds isolated from the kratom plant. Red colour-coded compound names are those phytocompounds with the most reported pharmacological properties.

(neuroprotective) (Lagunin et al., 2020), guattegaumerine (anticancer, neuroprotective) (Lü et al., 2009; Sergent et al., 2022), hirsuteine (calcium channel blocker, antitumour, neuroprotective) (Kawakami et al., 2011; Meng et al., 2023), huperzine B (neuroprotective, anti-inflammatory, antinociceptive, anticonvulsant) (Ferreira et al., 2016; Simunkova et al., 2019), ismine (antiparasitic) (Martinez-Peinado et al., 2022), isoliensinine (antidepressant, antitumour, cardioprotective, antioxidant, and anti-HIV) (Kashiwada et al., 2005; Cheng et al., 2021; Yao et al., 2023), isomitraphylline (anti-leukemic, immunostimulant) (Pilarski et al., 2007; Azevedo et al., 2019), isomitrafoline (auxiliary immunostimulant) (Hassan et al., 2013), isopteropodine (antileukemic, positive modulators for muscarinic M1 and 5-HT2 receptors) (Kang et al., 2002; Bacher et al., 2006), isorhynchophylline (antihypertensive, neuroprotective, anticancer, anticonvulsant) (Chen et al., 2014; Lee et al., 2017; Wang et al., 2022), isospeciofoleine (indole alkaloid), isospeciofoleine, isospeciofoline (analgesic, antitussive), L-pipecolic acid (laxative, antidepressant) (Li et al., 2023), mitrafoline (anti-leukemic, antihypertensive, antiamnesic), mitraphylline (antitumour, neuroprotective, potent muscle relaxant, vasodilator, diuretic) (Azevedo et al., 2018; Tan and An, 2020), lumichrome (anticancer) (Chantarawong et al., 2019), prosopinine (anaesthetic, antibiotic, analgesic, anti-neuroinflammatory) (Lai et al., 2022), rhynchophylline (neuroprotective, cardioprotective, calcium channel blocker) (Frank et al., 2014), salsoline (antibacterial) (Ibrayev et al., 2022), salsolinol (cardioprotective, neuroprotective) (Kurnik-Łucka et al., 2020), speciofoline (mild analgesic) (Jeng-Yeou Chear et al., 2021), speciophylline (anti-leukemic), spirolucidine, tetrahydroalstonine (hypoglycemic, anti-adrenergic), theophylline (renal protective) (Pacifici, 2015), tryptamine (psychosis) (Castelhano et al., 2021), vasicinol (anti-infertility), yohimbine (central nervous system stimulation, antidiuresis) (Jabir et al., 2022), 1,2,3,4-tetrahydro-β-carboline-1,3-dicarboxylic acid (anti-inflammatory, antioxidant) (Park et al., 2014), and 2-[4-(3,4-Methylenedioxyphenyl)butyl]-4(1H)-quinolinone (antifungal) (Oliva et al., 2003).

Despite the numerous alkaloids isolated and identified from kratom leaves, only four compounds – mitragynine, 7-hydroxymitragynine (7-HMG), corynantheidine, and speciociliatine – are known to be pharmacologically active. The primary constituent alkaloid in kratom, mitragynine, an indole alkaloid, was first isolated by Field in 1921 (Field, 1921) and later demonstrated to have partial agonist activity on opioid receptors. This white amorphous powder is soluble in several solvents, such as alcohol, chloroform, and acetic acid. It was fully characterised via X-ray crystallography in 1965 (Zacharias, Rosenstein, and Jeffrey, 1965).

In addition to pharmacologically active alkaloids, kratom contains several secondary metabolites, including saponins, iridoids, monoterpenoids, triterpenoids, and polyphenols (rutin, quercetin, quercitrin, isoquercitrin, apigenin, and apigenin 7-glycosides). Other metabolites present in kratom include (-)-alstolucine A, 11-methoxyvinorine (a crucial step in ajmaline biosynthesis) (Bayer, Ma and Stöckigt, 2004), 3-α(S)-Strictosidine (an intermediate product from the monoterpenoid indole alkaloid biosynthesis pathway) (Brose et al., 2021), perakine (ajmaline biosynthetic pathway) (Cai et al., 2019), cinegalline (quinolizidine alkaloid), kaempferol, caffeine, benzyl-β-d-glucopyranoside, quercetin-3-galactoside-7-rhamnoside, hyperoside, quinovic acid, 3-oxo-α-ionyl-O-β-d-glucopyranoside, etc.

12.5 BIOACTIVITIES OF KRATOM

Native plants are often used as essential sources of traditional human medicines to treat or alleviate symptoms of various diseases. In the local Thai community, *M. speciosa* has been recognised as a traditional medicine for several discomforts and diseases. Kratom has been shown to have a variety of therapeutic potentials, including anti-inflammatory, antimicrobial, antinociceptive, antidiabetic, anti-dyslipidaemic, and antioxidant activities in both *in vitro* and *in vivo* biological studies. These findings may provide scientific evidence for related traditional medicinal practices.

12.5.1 Analgesic Effects of Kratom

Kratom leaves have been reported to be commonly used for self-treatment of various conditions or ailments, including pain and tiredness. Therefore, recent studies on the analgesic properties of *M. speciosa* leaves provide a better understanding of their traditional uses.

The potential antinociceptive effect of different kratom extracts was primarily identified by observing the latency time of a rat model's response to high temperature and formalin-induced pain stimulation (Shaik Mossadeq et al., 2009; Sabetghadam, Ramanathan and Mansor, 2010). It was ascertained that the antinociceptive action of 20 mg/kg alkaloid, 200 mg/kg methanolic, and 400 mg/kg aqueous extracts was reversed by an opioid antagonist, naloxone (Sabetghadam, Ramanathan and Mansor, 2010). Another confirmed the analgesic effect of kratom in an *in vivo* murine model. The methanolic kratom extract was shown to inhibit formalin-induced pain in rats (Shaik Mossadeq et al., 2009).

Goh et al. (2021) studied the impact of kratom leaf extracts, including aqueous, methanol, ethyl acetate, accelerated solvent extraction (ASE) with ethanol, and ASE with ethyl acetate, on analgesic activity. All the extracts were subjected to evaluation of their cytotoxicity on HeLa, Chang liver and HEK-293 cell lines, demonstrating that all the kratom extracts except ASE with ethyl acetate extract did not affect the viability of the cells (Goh et al., 2021). The ASE of kratom ethanol extract demonstrated a better antinociceptive effect compared to methanol and ethyl acetate leaf extracts. Interestingly, both morphine-treated and 200 mg/kg ASE ethanol kratom extracts exerted comparable onset and duration of analgesic effects in the tail-flick test. In contrast, in the hot plate test, ASE ethanol leaf extract displayed a delayed onset of action by about 60 minutes and a shorter duration of antinociceptive effect. These experiments indicated that kratom leaf extract could alleviate pain in a mouse model (Goh et al., 2021).

Mitragynine, one of kratom's active compounds, was discovered to exhibit morphine-like effects or act as an agonist for opioid receptors in the guinea-pig ileum (Watanabe et al., 1997) and to have pain-relieving effects via action on supraspinal μ and δ opioid receptors (Matsumoto et al., 1996; Takayama, 2004). 7-Hydroxymitragynine was reported to have a high affinity for μ opioid receptors and to prevent the contraction induced by electrical stimulation through opioid receptors in the guinea-pig ileum (Watanabe et al., 1997; Vicknasingam et al., 2020). Additionally, a previous study reported that mitragynine can impede the contraction of smooth muscle induced by electrical stimulation in guinea pigs through the blockade of neuronal T- and L-type calcium (Ca2+) channels and inhibit the influx of Ca2+ in neuronal cells (Matsumoto et al., 2005).

In addition to mitragynine and 7-hydroxymitragynine, speciociliatine, paynantheine, speciogynine, specioliatine, 9-hydroxycorynantheidine, and corynantheidine from kratom were discovered to activate analgesic effects through μ opioid receptors (Kruegel et al., 2016; Obeng et al., 2020). On the other hand, flavonol groups from flavonoids have also been found to induce antinociceptive effects in murine models via central and peripheral mechanisms (Goh et al., 2021). For instance, quercetin has analgesic effects and can bind to the α2-adrenergic receptors of both central and peripheral pathways. Both quercetin and its derivative, rutin, can initiate the cyclic guanosine monophosphate/protein kinase-G (cGMP/PKG) ATP-sensitive potassium pathway and inactivate neuronal signalling to halt pain sensation (Ferraz et al., 2020).

The antinociceptive effect of kratom has been further demonstrated in human subject studies. Pain tolerance in humans was evaluated by the period between the onset of pain and the withdrawal of the hand from an ice bath. It was discovered that kratom users had significantly elevated pain tolerance (Vicknasingam et al., 2020), suggesting that kratom exhibited analgesic properties.

Other than the typical opioid receptor activation, mitragynine as a potential pain-alleviating agent is found to mediate its analgesic effect via anti-inflammatory and non-opioid receptor pain pathways, such as the transient receptor potential cation channel subfamily V member 1 (TRPV1) receptors (Mat et al., 2023). TRPV1 receptors are broadly distributed on C and Aδ afferent neurons and play a role as thermal-sensor receptors, making them potential therapeutic targets for inflammatory pain therapy (McGaraughty et al., 2009). A recent study reported that administration of mitragynine inhibits writhing effects and incurs low pain-related withdrawal behaviours in acetic acid-induced pain sensory of Sprague-Dawley rats, as well as lowers TRPV1 expression in the thalamus and somatosensory (SC) regions of the rat brain (Mat et al., 2023). It also shows that mitragynine from the kratom plant can decrease the transmission of pain stimuli from the pain-induced peripheral site to the brain, explicitly in the SC region where pain sensations are finally perceived (McGaraughty et al., 2009; Mat et al., 2023). All of these studies indicate that kratom has antinociceptive activities and support its folkloric use for pain treatments.

12.5.2 Antimicrobial Activity of Kratom

Kratom plants in Thailand are commonly used as a traditional remedy to relieve several food-borne diseases related to microbial infections, such as diarrhoea and stomach discomfort. Therefore, understanding the antimicrobial activities of kratom provides deeper insights into its traditional uses.

The leaf extracts of kratom have been used to test for various antimicrobial activities. The methanolic kratom leaf extract was assessed for antimicrobial activity and demonstrated moderate inhibition of Aeromonas hydrophilia growth, with an average kill zone diameter of 9.55 mm at 24% of kratom extract (Juanda, Andayani, and Maftuch, 2019). Next, the supplemented kratom leaf methanolic extract at a concentration of 25–100% was able to prevent the growth of *Escherichia coli* (*E. coli*) and *Streptococcus pneumoniae* (Salim et al., 2021). Based on an *in silico* study, it was deduced that a secondary metabolite called quercetin from kratom interacted with the ATP1 protein of microbes, thereby blocking bacterial growth via inhibition of oxidative phosphorylation in mitochondria (Wang et al., 2018; Abdul Rahman et al., 2022).

Furthermore, *M. speciosa* leaf alkaloid and methanolic extracts were screened and exhibited antimicrobial activity against *Salmonella typhi* and *Bacillus subtilis*. The inhibition zone of the methanolic extract was 29–30 mm and 30–33 mm for the alkaloid extract, at minimal inhibitory concentrations (MIC) of 6.25 mg/mL and 3.25 mg/mL, respectively (Parthasarathy et al., 2009). Besides, dichloromethane and hexane kratom extracts exhibited low-to-moderate antimicrobial activities against several bacteria strains. The dichloromethane extract inhibited the growth of *Bacillus subtilis* (*B. subtilis*), *Staphylococcus epidermis* (*S.* epidermidis), *Klebsiella pneumoniae* (*K. pneumoniae*), and *Pseudomonas aeruginosa* (*P. aeruginosa*) with zones of inhibition (ZI) of 16.5 mg/mL, 16.0 mg/mL, 12.0 mg/mL and 14.5 mg/mL, respectively, while the hexane kratom extract had ZIs of 15.5 mg/mL, 16.5 mg/mL, 16.0 mg/mL, and 15.0 mg/mL against *B. subtilis*, *S.* epidermidis, *E. coli* and *P. aeruginosa* (Abdul Rahman et al., 2022).

Moreover, both pure kratom leaf extract and fermented kratom leaf extract with *Lactobacillus rhamnosus* displayed significant antibacterial activity against the biofilm formation of *Staphylococcus aureus*, methicillin-resistant *Staphylococcus aureus* (MRSA), *E. coli*, and *E. coli* O157:H7 with MIC values of 50 mg/mL each. However, kratom leaf extract fermented with probiotics exhibited more potent inhibition towards pathogens due to the growth of probiotics during fermentation, which stimulated and enhanced the antimicrobial activity of kratom leaf extract (Sornsenee, Chimplee and Romyasamit, 2023).

12.5.3 Anti-inflammatory Activity of of Kratom

The inflammatory microenvironment is a hallmark of various diseases, for example, diabetes, arthritis, and rheumatoid disorders, related to chronic inflammation (Meizlish et al., 2021; Baechle et al., 2023). Several kratom species in Thailand have been used to treat inflammatory disorders as a traditional regimen, which suggests that the phytochemical compounds in the plant might possess anti-inflammatory activities.

The methanolic *M. speciosa* extract at doses of 100–200 mg/kg was shown to have anti-inflammatory activity by halting the progression of carrageenan-stimulated oedema and reducing the formation of granulomatous tissue in a rat's paw (Shaik Mossadeq et al., 2009). Both pure kratom leaf extract and fermented kratom leaf extract with *L. rhamnosus* were tested for their anti-inflammatory activities. As nitric oxide (NO) production is an indicator of immune response to inflammation, the effects of both pure kratom leaf extract and fermented kratom leaf extract on NO production in lipopolysaccharide (LPS)-stimulated murine RAW 264.7 macrophage cells were determined. The results revealed that all these treatments significantly inhibited NO production, ranging from 43.93 ± 8.04 μM to 51.93 ± 15.35 μM, without affecting the viability of RAW 264.7 macrophage cells (Sornsenee, Chimplee and Romyasamit, 2023). The reduction in NO levels in both kratom extract treatments was similar to that of the anti-inflammatory drug aspirin, which was at 48.76 ± 3.69 μM. The kratom extract treatments also reduced the levels of the pro-inflammatory cytokines tumour necrotic factor-α (TNF-α), IL-1β, and IL-6 in LPS-induced RAW 264.7 cells (Sornsenee, Chimplee and Romyasamit, 2023).

As for the mitragynine compound from kratom, it was discovered to inhibit the LPS-induced cyclooxygenases (COX-1 and COX-2) and reduce the production of prostaglandin (PGE2) in RAW264.7 macrophage cells. Mitragynine can attenuate the inflammatory response in macrophage cells, proposing that mitragynine could be a potential anti-inflammatory agent (Utar et al., 2011). The ethanolic kratom extract also exhibited similar anti-inflammatory activity. The extract showed low cytotoxicity at an IC_{50} value of 1 mg/mL towards skin cells like fibroblasts, keratinocytes, melanocytes, and macrophages. Upon induction with LPS, the genetic expression of COX-2, IL-1β, IL-6, and inducible nitric oxide synthase (iNOS) was suppressed significantly in *M. speciosa* ethanolic extract-treated macrophages. Mitragynine from the ethanolic extract was found to be permeable on porcine ear skin with the aid of hydrophilic chemical enhancers, suggesting that mitragynine could be used for transdermal and/or topical wound products in the future (Tuntiyasawasdikul et al., 2024).

Apart from mitragynine, epicatechin is present in low amounts in the kratom plant. This compound exhibits excellent anti-inflammatory effects and prevents certain chronic conditions, including metabolic diseases and certain cancers (Shay et al., 2015). For neurodegenerative disorders, it is speculated that kratom can confer neuroprotection via the inhibition of COX-2 activity (Hossain et al., 2023). Amyloid beta (Aβ) peptides are derived from β-Amyloid precursor protein (APP) by enzymatic cleavage using (α, β, and γ) secretases and form the Aβ plaque. The aggregation of Aβ1-42 forms plaques that trigger calpain and downregulate p35 into p25 (Caselli, Knopman and Bu, 2020). This hyperactivates CDK5, causing tau protein to be hyperphosphorylated and initiating the development of neurofibrillary tangles (NFTs). By promoting the upregulation of nuclear factor-kappa B (NF-κB) and adaptor protein 1 (AP-1) transcription factors, followed by the generation of pro-inflammatory cytokines such as COX-2, TNF-α, NO, PGE2, IL-1, IL-6, and IL-7, which harm cholinergic neurons, β-amyloid plaques consequently induce neurotoxicity or activation of microglia (N. M. Li et al., 2019; Caselli, Knopman and Bu, 2020). The astrocytes directly stimulated by these pro-inflammatory cytokines also elevate inflammatory signals, ultimately resulting in neuroinflammation and neurodegeneration. The anti-inflammatory properties of kratom are believed to improve immunity, tissue repair, and healing, as well as inhibition of the production of pro-inflammatory mediators and vascular permeability effect (Shaik Mossadeq et al., 2009). The overall anti-inflammatory activity activated by kratom is summarised in Figure 12.3.

12.5.4 Antidiabetic and Anti-dyslipidaemia Effects of Kratom

Modern biological studies have revealed that *M. speciosa* also displays antidiabetic properties, even though traditional practices in the Thai community may be limited. Therefore, current biological research could complement the depth and breadth of the ethnomedicinal uses of kratom.

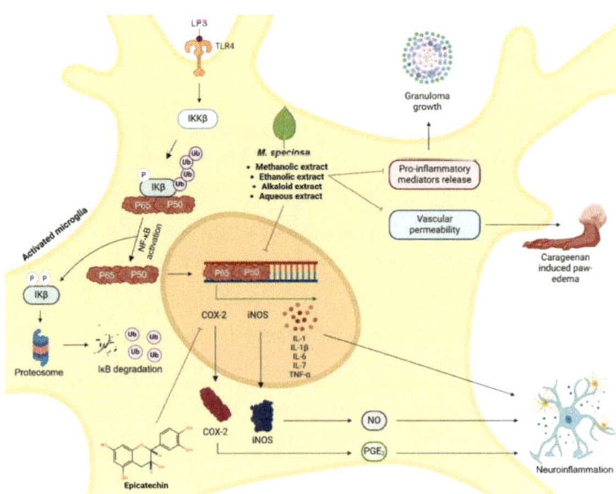

FIGURE 12.3 The summary of kratom-induced anti-inflammatory effect on microglia (created with BioRender.com).

Several studies have found that kratom users in Thailand exhibited antidiabetic activity. For instance, the 1000 μg/mL ethanolic extract of kratom leaf was able to inhibit the activity of the α-glucosidase enzyme with IC50 values of 382.64 ± 16.18 μg/mL and 409.43 ± 19.14 μg/mL (Pipatrattanaseree et al., 2023), demonstrating the antidiabetic potential of kratom. Another study by Zhang et al. (2023) also showed that the ethanolic kratom extract had potent inhibition on the *in vitro* α-glucosidase enzyme at an IC_{50} value of 128.96 ± 4.75 μg/mL. Interestingly, they carried out the study on the effect of kratom extract in fructose and streptozocin type II diabetes (T2D) rats, showing that the condition of T2D rats significantly restored blood glucose levels, improved insulin secretion, restored hepatorenal biomarker alterations (alanine transaminase, aspartate transaminase, alanine phosphatase, creatinine, and blood urea nitrogen), as well as reconditioned pancreatic histological anomalies and immunohistochemical aberrations (Zhang et al., 2023). Elevated reactive oxygen species (ROS) and oxidative stress are some of the hallmarks of hyperglycaemia (Oguntibeju, 2019). After the administration of ethanolic kratom leaf extract in T2D rats, the oxidative stress and ROS were significantly reduced due to an increment of superoxide dismutase (SOD) activity and glutathione (GSH) levels with a corresponding decline in malondialdehyde (MDA) levels in rat serum (Zhang et al., 2023). The potency of kratom as an antidiabetic agent suggests it could serve as an alternative remedy in the treatment of diabetes and its associated comorbidities.

The inhibitory activity of aqueous kratom extracts on angiotensin-I converting enzyme (ACE) was performed to investigate the potential of kratom as an anti-hypertensive agent. It was revealed that 1056.08 μg/mL of aqueous kratom extract was required to inhibit the activity of ACE (Pipatrattanaseree et al., 2023). Additionally, ethanolic kratom leaf extract was also shown to have the ability to reduce T2D rats' serum cholesterol, triglyceride, and low-density lipoprotein (LDL) concentrations while increasing the high-density lipoprotein (HDL) level (Zhang et al., 2023). This study is in parallel with a human subject study executed by La-Up, Saengow and Aramrattana (2021) in Nam Phu subdistrict, Surat Thani province, Thailand, where the usage of kratom was able to increase HDL levels (>60 mg/dL) while reducing triglyceride levels (<90 mg/dL) and waist circumferences. A previous study performed in Malaysia also showed a similar trend, in which HDL levels in kratom users were higher than in non-kratom users (Singh, Müller et al., 2018). These studies provided evidence of kratom use and favourable lipid profiles in humans, suggesting the potential pharmaceutical application of kratom in the prevention of coronary heart disease or cerebrovascular diseases (La-Up, Saengow, and Aramrattana, 2021).

The proposed underlying mechanism of how *M. speciosa* prevents metabolic diseases is that mitragynine inhibits the expression of COX-2, as COX-2 is known to be positively correlated with serum triglycerides. Mitragynine can initiate the activation of opioid receptors (Jeng-Yeou Chear et al., 2021). Opioid receptors have vital functions in lipid and energy metabolism and can affect the regulation of glucose homeostasis in living organisms (Shang et al., 2015). Therefore, kratom could reduce the occurrence of metabolic disorders possibly caused by the effect of mitragynine on the expression of COX-2 and opioid receptors.

12.5.5 Antioxidant Effects of Kratom

Various acute and long-term illnesses in humans, including inflammatory diseases, cardiovascular diseases, and malignancies, are caused by different levels of oxidative stress (Lü et al., 2009; Oguntibeju, 2019; Simunkova et al., 2019). In turn, the ethnomedicinal application of Thai medicinal kratom can be justified by its antioxidant properties to a certain extent.

Different extracts from *M. speciosa* have been examined for their antioxidant effects by various assays such as the total phenolic content assay, total flavonoid content assay, Dragendorff reagent assay, Wagner reagent assay, vitamin C antioxidant activity assay, ferric reducing antioxidant power (FRAP) assay, 2,2-diphenyl-1-picrylhydrazyl (DPPH), and 2,2′ -azino-bis-3-ethylbenzthiazoline-6-sulfonic acid (ABTS) free radical scavenging activity assays (Yuniarti et al., 2020; Abdul Rahman et al., 2022; Zailan, Sarchio and Hassan, 2022; Sornsenee, Chimplee and Romyasamit, 2023; Zhang et al., 2023).

To evaluate the antioxidant properties of the kratom leaves' aqueous, methanolic, and alkaloid extracts, TPC, TFC, and DPPH radical scavenging assays were carried out. The TPC of the water, alkaloid, and methanolic leaf extracts was 66.0, 88.4, and 105.6 mg GAE/g, while the TFC was 28.2, 20.0 and 91.1 mg CAE/g, respectively. As for the DPPH IC50 values, the aqueous, alkaloid, and methanolic kratom leaf extracts were 213.4, 104.81, and 37.08 µg/mL, respectively. It was suggested that antioxidant activities were correlated with the TPC, implying that the presence of high phenolic content in the methanolic extract contributed to the high antioxidant activity (Parthasarathy et al., 2009).

In addition, the ethanolic kratom leaf extract demonstrated scavenging solid activity against DPPH with an IC_{50} value of 38.56 µg/mL, indicating that kratom possesses potent antioxidant activity (Yuniarti et al., 2020). This result agrees with another study of ethanolic kratom leaf extracts by Zhang et al. (2023). The ethanolic extract was evaluated for its phenolic composition and antioxidant activity. It was shown that the TPC was 24.62 mg GAE/g, whereas the antioxidant activity of ethanolic leaf extract (IC50 value) in the DPPH, ABTS, and FRAP assays were 21.61 ± 0.01 µg/mL, 9.33 ± 0.01 µg/mL, and 89.31 ± 1.44 mg of TE/g of sample, respectively (Zhang et al., 2023). Kratom leaf extract and fermented kratom leaf extract with *L. rhamnosus* possessed high TPC of 48.34 – 53.25 mg GAE/g, moderate TFC ranging from 3.62–4.05 mg QE/g, high DPPH scavenging percentage of 93.13%–94.33%, and high ABTS scavenging activity of 99.35%–99.94% (Sornsenee, Chimplee and Romyasamit, 2023). Overall, all the kratom extracts, as mentioned earlier, have antioxidant effects and inhibit the expression of pro-inflammatory cytokines.

12.5.6 Implication of Kratom on the Neuronal System

One of the most abused plant-based psychotropic drug sources, kratom, contains a potent psychoactive compound called mitragynine, which causes neuroplasticity in the brain's reward system and exhibits behavioural effects akin to those of opioids (Sethi et al., 2020). Chronic administration of kratom has been connected to cognitive deficits. Significant advancements in pain management have been made possible by multimodal analgesic techniques, including improved tolerance and enhanced efficacy of opioids. Combining analgesics with various modes of action has been shown to effectively treat multiple pain problems while reducing side effects and optimising the benefits of the additive and synergistic actions of the separate drugs (Ferraz et al., 2020).

The most prevalent and potent indole alkaloid in kratom, mitragynine, is speculated to have a role in modulating the Keap-1/Nuclear factor-erythroid factor 2-related factor 2 (Keap-1/Nrf2) pathway to confer neuroprotection (Hossain et al., 2023). The adrenergic, serotonergic, and opioid-like actions of mitragynine are mediated by mechanisms that differ structurally and pharmacologically from those of conventional opioids (Kruegel et al., 2016). Apart from mitragynine and several other alkaloids in this group, the majority of them might be crucial in regulating the excessive generation of ROS within neuronal cells, leading to neuronal cell death through the exposure of hydrogen peroxide (H_2O_2). It has been demonstrated that the nuclear translocation of Nrf2 is mainly responsible for the activation of antioxidative genes of Nrf2, including heme oxygenase-1 (HO-1) and NAD(P)H quinone dehydrogenase 1 (NQO1) (Hossain et al., 2023). The mechanism of action is summarised in Figure 12.4.

Kratom was demonstrated to increase manual laborers' alertness (Singh et al., 2019). Hence, various *in vitro* inhibitory activities on neurogenic-related enzymes were performed. The ethanolic kratom extracts showed highly potent inhibition of both acetylcholinesterase (AChE) and butyrylcholinesterase (BChE), with IC50 of 127.99 ± 6.15 μg/mL and 64.40 ± 16.74 μg/mL, respectively (Pipatrattanaseree et al., 2023), proposing that kratom extracts may serve as an alternative natural product

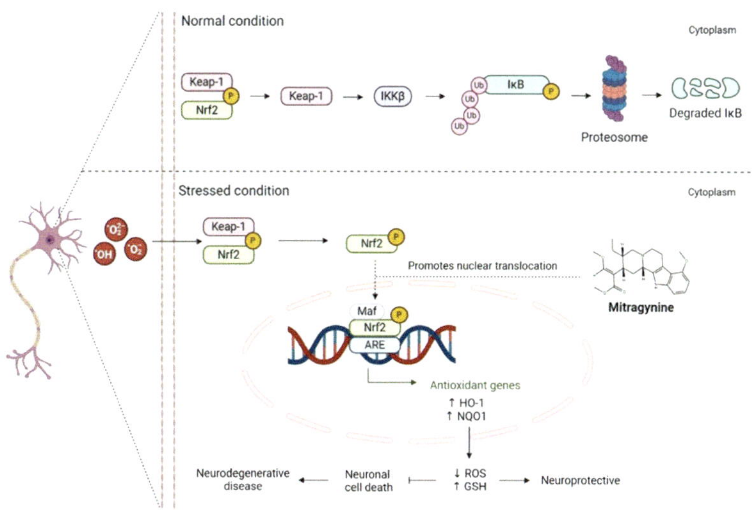

FIGURE 12.4 Cascade mechanism of the Keap-1/Nrf2 associated pathway on the neuroprotective effects of mitragynine (Created with BioRender.com).

to improve the condition of neurological-related diseases like Alzheimer's disease (AD) and Parkinson's disease (PD). Apart from AChE and BChE, the suppression of the activities of adenosine triphosphatase (ATPase), adenosine diphosphatase (ADPase), and monoamine oxidase (MAO) are proposed to be effective for neurodegenerative disorders. Superoxide dismutase (SOD) and catalase (CAT) are crucial antioxidant enzymes that function to halt the production of harmful H_2O_2 (Wang et al., 2020). These enzymes were revealed to be significantly elevated in the brain tissue upon administration of kratom hydroalcoholic extract (Chen, Fei, and Olatunji, 2022). Increased Aβ oligomer levels have been reported to facilitate the progression of oxidative stress, synaptic loss, neuroinflammation, and neuronal cell death. A recent study demonstrated mitragynine and mitragynine oxidole B in kratom leaves to prevent the AD-associated AChE enzyme (Innok et al., 2021). The mechanism of action of the neuroinflammation inhibition of kratom is summarised in Figure 12.5.

ROS are predominantly responsible for the abnormal Aβ peptide aggregation in the brain, which promotes the disease progression of AD (Hensel, Rademacher, and Claus, 2015). Isorhynchophylline, another alkaloid in kratom, is helpful in the treatment of AD. This is mainly due to isorhynchophylline's neuroprotective traits, whereby this phytocompound acts by mediating ROS removal and reducing the Bcl-2/Bax gene expression levels, leading to a decrement of Aβ-induced neuronal cell death in the hippocampus (H. Q. Li et al., 2019).

Kratom has been used to deal with drug and alcohol withdrawal symptoms such as anxiety, mood swings, and depression (Kumarnsit, Keawpradub and Nuankaew, 2007; Embry et al., 2009; Wearne and Cornish, 2018). Hitherto, kratom alkaloid extract was discovered to exhibit similar local field potential (LFP) patterns of electrical brain waves and mitigate alcohol withdrawal symptoms as the standard antidepressant drug fluoxetine (Cheaha et al., 2015). Aqueous kratom extract, too, attenuated ethanol withdrawal-stimulated depressive behaviour, possibly via its antidepressant-like activity.

In addition, the hydroalcoholic extract of kratom was also shown to have positive impacts on diabetes-induced anxiety and depression in T2D rats. For instance, the diabetic rats with kratom treatment exhibited attenuated the depressive and anxiety-like behavioural deficits. At the same time, the brain oxidative stress and pro-inflammatory cytokine levels such as IL-1β, IL-6, and TNF-α were alleviated (Chen, Fei, and Olatunji, 2022). However, kratom is like a double-edged sword; it could be beneficial for health, or it may induce detrimental effects for long-term users. One of the studies stated that the exposure of kratom leaf's alkaloid, named mitragynine, and lyophilised kratom decoction (LKD) to adolescent Sprague-Dawley rats caused the impairment of long-term object recognition memory. Both mitragynine and LKD altered the metabolite profiles of the brain, which may be responsible for the cognitive behavioural effects, indicating that the adolescent brain is susceptible to the influence of early kratom use (Suhaimi et al., 2023). In contrast, the consumption of kratom tea by regular kratom users above 18 years old did not affect motor, memory, attentional processing, and executive function (Singh et al., 2019). This could be because kratom users have not consumed kratom from a young age.

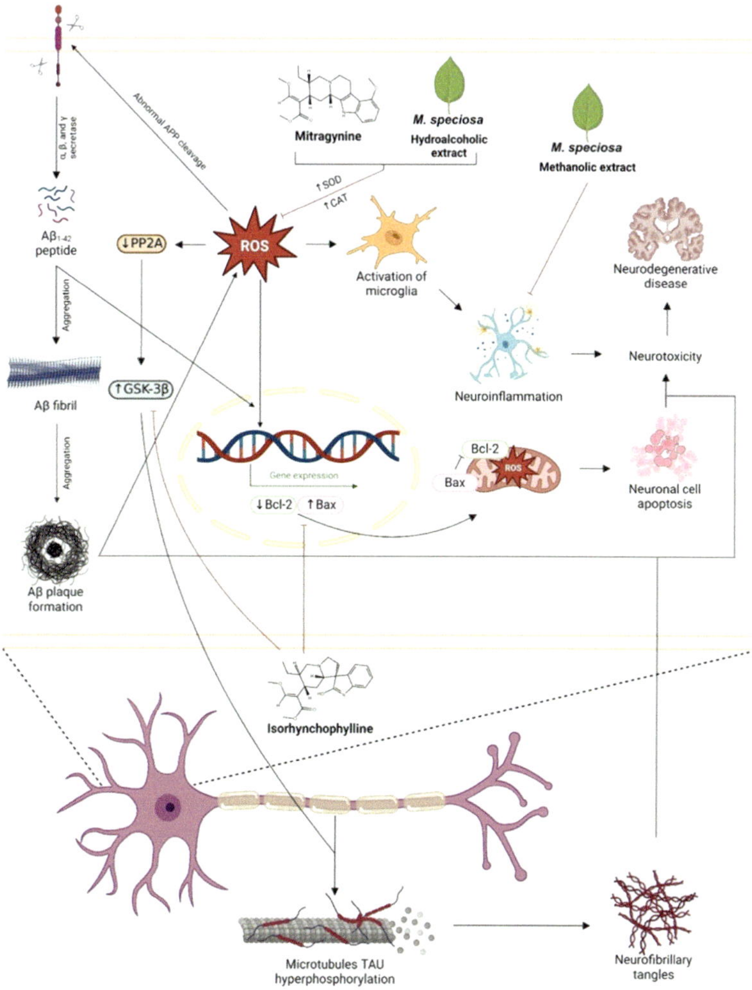

FIGURE 12.5 Mechanism of action on gene regulation by *M. speciosa* to halt the development of neuroinflammation (Created with BioRender.com).

As aforementioned, the kratom plant can act on opioid receptors like morphine, heroin, and methamphetamine (Singh et al., 2022). For instance, mitragynine and speciociliatine in *M. speciosa* could confer DNA protection abilities that are 20- to 1200-fold greater than morphine. These alkaloids are thought to directly or indirectly affect the neuroprotective effects of alkaloids found in alkaloid-rich plant products like kratom. The primary alkaloid derived from kratom, mitragynine, was given at doses ranging from 0.5 to 20 g/mL in a dose-dependent manner. Therefore, the LPS-induced mRNA expression of COX-2 was significantly inhibited. This was followed by a reduction in PGE2 synthesis, suggesting the anti-inflammatory properties of mitragynine in kratom (Utar et al., 2011). All these alkaloids found in alkaloid-rich plant products like kratom are thought to have a direct or indirect effect on neuroprotective effects.

12.5.7 Anticancer Effects of Kratom

Kratom has demonstrated anti-tumour activities, with dichloromethane kratom extract showing superior cytotoxicity activity against various cancer cell lines compared to hexane kratom extract. Dichloromethane kratom extract displayed a potent antiproliferative effect against breast cancer (A549, MCF-7, and MDA-MB-231) and colon cancer (SW948 and HT29) cell lines with IC50 values ranging from 11.76 μg/mL to 31.52 μg/mL. In contrast, hexane kratom extract only displayed weak anticancer properties, with an IC50 of 92.62 μg/mL against SW98 cells (Abdul Rahman et al., 2022).

12.5.8 Gastrointestinal Protective Effect

The bidirectional gut-brain axis links the gastrointestinal tract and the brain, enhancing communication between the stomach and the central nervous system. This interaction affects diverse physiological processes, emotions, and behavioural patterns. The relationship between the gut microbiota, hormones, immunological responses, and neural pathways, and immunological responses highlights the importance of gut health for overall brain function and mental health (Arneth, 2018).

In a study on adult Wistar rats with castor oil–induced diarrhoea, 200 mg/kg and 400 mg/kg doses of ethanolic kratom leaf extracts were administered. The extracts inhibited intestinal transit and exhibited antidiarrheal effects on the rat gastrointestinal tract. Additionally, a decrease in the body weight of the rats was observed, possibly indicating action by mitragynine on pathways other than opioid receptors (Chittrakarn et al., 2008).

In another study, mitragynine suppressed the release of gastric acid stimulated by 2-deoxy-D-glucose in a dose-dependent manner upon administration. The effect of mitragynine was reversed by naloxone, suggesting the involvement of opioid receptors (Tsuchiya et al., 2002). However, the basal gastric acid secretion into the lateral ventricle remained unaffected by the central injection of mitragynine. Besides its implications on anorexia and weight loss, mitragynine also directly hindered the activation of neurons in the lateral hypothalamus (Tsuchiya et al., 2002). Furthermore, 7-hydroxy mitragynine from kratom predominantly acts on μ-opioid receptors in a dose-dependent manner and inhibits gastrointestinal transit in mice at an effective dosage of 1.19 mg/kg (Matsumoto, Hatori, et al., 2006).

12.6 DRUG ABUSE

Kratom offers various benefits, as mentioned earlier. However, its usage has been associated with dependence, adverse effects, and withdrawal symptoms. For example, regular kratom users show increased expression of endoplasmic reticulum (ER) stress sensor mRNA, including binding immunoglobulin protein (BiP), activating transcription factor 4 (ATF4), and CCAAT-enhanced binding protein homologous protein (CHOP) in peripheral leukocytes. The high expression of these mRNAs is speculated to be associated with the severity of kratom dependence and depressive symptoms (Fauzi et al., 2022). Besides, a recent study also demonstrated that regular kratom users had exacerbated levels of ER stress sensors, namely BiP, spliced X-box-binding protein 1 (sXBP1), ATF4, CHOP, and c-Jun N-terminal kinase (JNK), resulting in dysregulation of the unfolded protein response to restore ER homeostasis, ultimately leading to cellular apoptosis (Yang et al., 2023). Figure 12.6 extensively illustrates the pathways involved in kratom dependence, suggesting that the upregulation of these proteins is correlated with kratom abuse disorder.

The safety and efficacy of kratom use are highly provocative. Since mitragynine and 7-hydroxy mitragynine in kratom bind non-selectively to different proteins, including G-protein receptors and cytochrome P450 (CYP) enzymes, adverse interactions may occur when kratom is co-ingested with drugs that bind to these targets (Todd et al., 2020). For instance, mitragynine induces cellular cytotoxicity and arrests the cell cycle at the G1/S phase in the human lymphoblastoid TK+/-MCL-5 cell line, proposing that CYP2E1 and 2A6 are involved in apoptosis. This also implies that individuals with high CYP2E1 activity, such as alcohol consumers, are more prone to *M. speciosa* and mitragynine toxicity (Saidin et al., 2015). Moreover, there have been lethal incidents related to intoxication associated with the usage of "Krypton," a mixture of mitragynine and O-desmethyl-tramadol (Arndt et al., 2011; Kronstrand et al., 2011). The O-desmethyl-tramadol addition to the dried leaves of kratom, however, has been implicated in these fatalities (Kronstrand et al., 2011). Given the significant consumption of kratom, further research in the fields of toxicology, pharmacology, socio-anthropology, and clinical medicine is required to address these understudied concerns.

12.7 CONCLUSIONS

Despite the negative and positive effects of *M. speciosa*, recent studies highlight its potential as a plant that offers analgesic, antimicrobial, anticancer, antidiabetic, anti-dyslipidaemic, antioxidant, and anti-inflammatory properties, as well as gastrointestinal and neuroprotective advantages. These diverse benefits undoubtedly attract significant attention.

To ensure that the benefits of this plant outweigh the harmful effects, such as abuse and dependence, further investigations should focus on all isolated compounds. Most of the phytocompounds from *M. speciosa,* except mitragynine and 7-hydroxymitragynine, have not been fully characterised and studied in terms of their biological activities. Therefore, more in-depth and thorough dose-dependent preclinical and clinical research is necessary to determine the optimum therapeutic windows for each disorder and reduce dependence incidents.

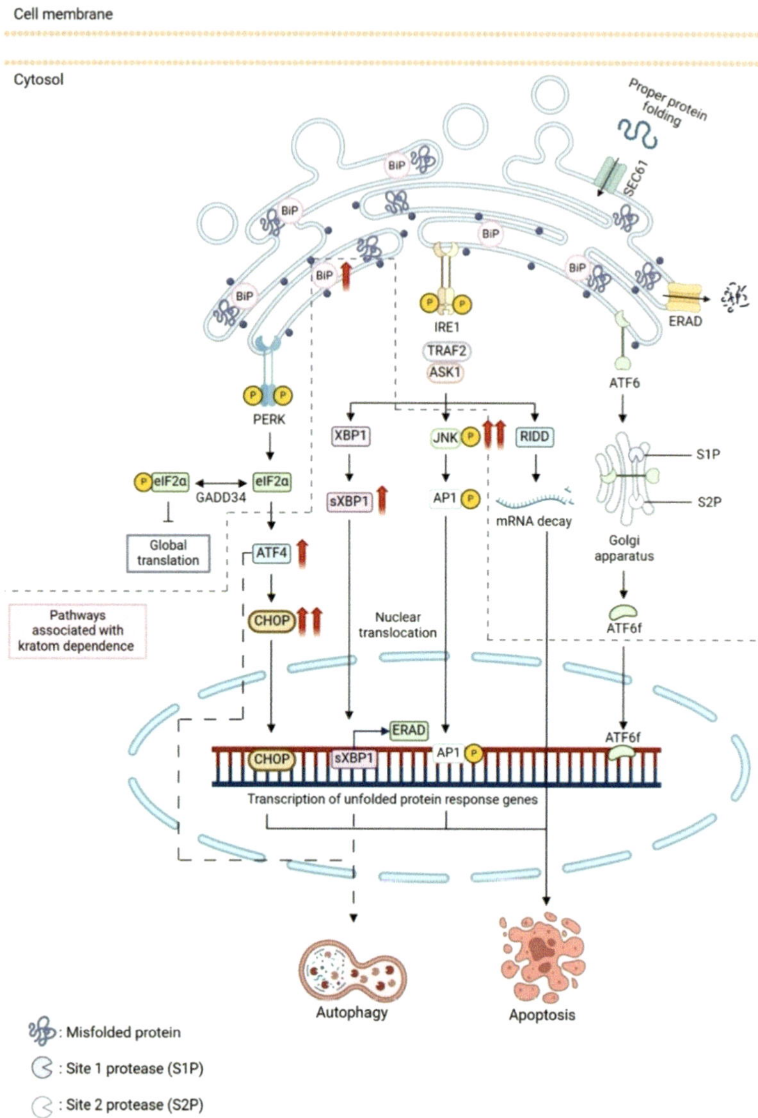

FIGURE 12.6 Kratom dependence pathways (Created with BioRender.com).

REFERENCES

Abdul Rahman, M.N. et al. (2022) 'Antioxidant, antimicrobial, and cytotoxic activities of the hexane and dichloromethane extracts of Malaysian Mitragyna spesioca Korth. leaves', *Malaysian Journal of Chemistry*, 24(2). Available at: https://doi.org/10.55373/mjchem.v24i2.191.

Agrawal, S.S. et al. (2011) 'Cytotoxic and antitumor effects of brucine on Ehrlich ascites tumor and human cancer cell line', *Life Sciences*, 89(5–6). Available at: https://doi.org/10.1016/j.lfs.2011.05.020.

Arndt, T. et al. (2011) 'Kratom alkaloids and O-desmethyltramadol in urine of a "Krypton" herbal mixture consumer', *Forensic Science International*, 208(1–3). Available at: https://doi.org/10.1016/j.forsciint.2010.10.025.

Arneth, B.M. (2018) 'Gut-brain axis biochemical signalling from the gastrointestinal tract to the central nervous system: Gut dysbiosis and altered brain function', *Postgraduate Medical Journal*. Available at: https://doi.org/10.1136/postgradmedj-2017-135424.

Azevedo, B.C. et al. (2018) 'Aqueous extracts from Uncaria tomentosa (Willd. ex Schult.) DC. reduce bronchial hyperresponsiveness and inflammation in a murine model of asthma', *Journal of Ethnopharmacology*, 218. Available at: https://doi.org/10.1016/j.jep.2018.02.013.

Azevedo, B.C. et al. (2019) 'Antioxidant activity of an aqueous leaf extract from uncaria tomentosa and its major alkaloids mitraphylline and isomitraphylline in caenorhabditis elegans', *Molecules*, 24(18). Available at: https://doi.org/10.3390/molecules24183299.

Bacher, N. et al. (2006) 'Oxindole alkaloids from Uncaria tomentosa induce apoptosis in proliferating, G0/G1-arrested and bcl-2-expressing acute lymphoblastic leukaemia cells', *British Journal of Haematology*, 132(5). Available at: https://doi.org/10.1111/j.1365-2141.2005.05907.x.

Baechle, J.J. et al. (2023) 'Chronic inflammation and the hallmarks of aging', *Molecular Metabolism*. Available at: https://doi.org/10.1016/j.molmet.2023.101755.

Bayer, A., Ma, X. and Stöckigt, J. (2004) 'Acetyltransfer in natural product biosynthesis – Functional cloning and molecular analysis of vinorine synthase', *Bioorganic and Medicinal Chemistry*, 12(10). Available at: https://doi.org/10.1016/j.bmc.2004.02.029.

Brose, J. et al. (2021) 'The Mitragyna speciosa (Kratom) genome: A resource for data-mining potent pharmaceuticals that impact human health', *G3: Genes, Genomes, Genetics*, 11(4). Available at: https://doi.org/10.1093/g3journal/jkab058.

Cai, S. et al. (2019) 'Enantioselective reduction of α,β-unsaturated ketones and aryl ketones by Perakine reductase', *Organic Letters*, 21(12). Available at: https://doi.org/10.1021/acs.orglett.9b00950.

Caselli, R.J., Knopman, D.S. and Bu, G. (2020) 'An agnostic reevaluation of the amyloid cascade hypothesis of Alzheimer's disease pathogenesis: The role of APP homeostasis', *Alzheimer's and Dementia*, 16(11). Available at: https://doi.org/10.1002/alz.12124.

Castelhano, J. et al. (2021) 'The effects of tryptamine psychedelics in the brain: A meta-analysis of functional and review of molecular imaging studies', *Frontiers in Pharmacology*. Available at: https://doi.org/10.3389/fphar.2021.739053.

Chantarawong, W. et al. (2019) 'Lumichrome inhibits human lung cancer cell growth and induces apoptosis via a p53-dependent mechanism', *Nutrition and Cancer*, 71(8). Available at: https://doi.org/10.1080/01635581.2019.1610183.

Cheaha, D. et al. (2015) 'Effects of an alkaloid-rich extract from Mitragyna speciosa leaves and fluoxetine on sleep profiles, EEG spectral frequency and ethanol withdrawal symptoms in rats', *Phytomedicine*, 22(11), pp. 1000–1008. Available at: https://doi.org/10.1016/J.PHYMED.2015.07.008.

Chen, F. et al. (2014) 'Urinary metabolites of isorhynchophylline in rats and their neuroprotective activities in the HT22 cell assay', *Fitoterapia*, 97. Available at: https://doi.org/10.1016/j.fitote.2014.05.022.

Chen, L., Fei, S. and Olatunji, O.J. (2022) 'LC/ESI/TOF-MS characterization, anxiolytic and antidepressant-like effects of Mitragyna speciosa Korth extract in diabetic rats', *Molecules*, 27(7). Available at: https://doi.org/10.3390/molecules27072208.

Cheng, Y. et al. (2021) 'Isoliensinine: A natural compound with "Drug-Like" potential', *Frontiers in Pharmacology*. Available at: https://doi.org/10.3389/fphar.2021.630385.

Chittrakarn, S. et al. (2008) 'Inhibitory effects of kratom leaf extract (Mitragyna speciosa Korth.) on the rat gastrointestinal tract', *Journal of Ethnopharmacology*, 116(1). Available at: https://doi.org/10.1016/j.jep.2007.11.032.

Creed, S.M. et al. (2021) 'Isolation and pharmacological characterization of six Opioidergic Picralima nitida alkaloids', *Journal of Natural Products*, 84(1). Available at: https://doi.org/10.1021/acs.jnatprod.0c01036.

Embry, D. et al. (2009) 'Behavioral and social correlates of methamphetamine use in a population-based sample of early and later adolescents', *Addictive Behaviors*, 34(4), pp. 343–351. Available at: https://doi.org/10.1016/J.ADDBEH.2008.11.019.

Fauzi, N.A.M. et al. (2022) 'Regular Kratom (Mitragyna speciosa Korth.) use and its association with endoplasmic reticulum stress response', *Journal of Addiction Medicine*, 16(6). Available at: https://doi.org/10.1097/ADM.0000000000000988.

Ferraz, C.R. et al. (2020) 'Therapeutic potential of flavonoids in pain and inflammation: Mechanisms of action, preclinical and clinical data, and pharmaceutical development', *Molecules*. Available at: https://doi.org/10.3390/molecules25030762.

Ferreira, A. et al. (2016) 'Huperzine A from Huperzia serrata: A review of its sources, chemistry, pharmacology and toxicology', *Phytochemistry Reviews*. Available at: https://doi.org/10.1007/s11101-014-9384-y.

Field, E. (1921) 'XCVIII. – Mitragynine and mitraversine, two new alkaloids from species of mitragyne', *Journal of the Chemical Society, Transactions*, 119. Available at: https://doi.org/10.1039/CT9211900887.

Firmansyah, A., Sundalian, M. and Taufiq, M. (2021) 'Kratom (Mitragyna speciosa korth) for a new medicinal: A review of pharmacological and compound analysis', *Biointerface Research in Applied Chemistry*. Available at: https://doi.org/10.33263/BRIAC112.97049718.

Frank, É. et al. (2014) 'Synthesis of novel steroidal 16-spiroisoxazolines by 1,3-dipolar cycloaddition, and an evaluation of their antiproliferative activities in vitro', *Molecular Diversity*, 18(3). Available at: https://doi.org/10.1007/s11030-014-9516-8.

García-Díez, E. et al. (2024) '(-)-Epicatechin and colonic metabolite 2,3-dihydroxybenzoic acid, alone or in combination with metformin, protect cardiomyocytes from high glucose/high palmitic acid-induced damage by regulating redox status, apoptosis and autophagy', *Food & Function*, 15(5), pp. 2536–2549. Available at: https://doi.org/10.1039/D3FO04039A.

Goh, Y.S. et al. (2021) 'Accelerated solvent extractions (Ase) of mitragyna speciosa korth. (kratom) leaves: Evaluation of its cytotoxicity and antinociceptive activity', *Molecules*, 26(12). Available at: https://doi.org/10.3390/molecules26123704.

Hassan, Z. et al. (2013) 'From Kratom to mitragynine and its derivatives: Physiological and behavioural effects related to use, abuse, and addiction', *Neuroscience and Biobehavioral Reviews*. Available at: https://doi.org/10.1016/j.neubiorev.2012.11.012.

Hensel, N., Rademacher, S. and Claus, P. (2015) 'Chatting with the neighbors: Crosstalk between Rho-kinase (ROCK) and other signaling pathways for treatment of neurological disorders', *Frontiers in Neuroscience*. Available at: https://doi.org/10.3389/fnins.2015.00198.

Hossain, R. et al. (2023) 'A critical review of the neuropharmacological effects of Kratom: An insight from the functional array of identified natural compounds', *Molecules*. Available at: https://doi.org/10.3390/molecules28217372.

Ibrayev, M.K. et al. (2022) 'Synthesis, structure and molecular docking of new 4,5-Dihydrothiazole derivatives based on 3,5-Dimethylpyrazole and cytisine and salsoline alkaloids', *Molecules*, 27(21), p. 7598. Available at: https://doi.org/10.3390/MOLECULES27217598/S1.

Innok, W. et al. (2021) 'In silico and in vitro anti-AChE activity investigations of constituents from Mytragyna speciosa for Alzheimer's disease treatment', *Journal of Computer-Aided Molecular Design*, 35(3). Available at: https://doi.org/10.1007/s10822-020-00372-4.

Jabir, N.R. et al. (2022) 'A literature perspective on the pharmacological applications of yohimbine', *Annals of Medicine*. Available at: https://doi.org/10.1080/07853890.2022.2131330.

Jeng-Yeou Chear, N. et al. (2021) 'Exploring the chemistry of alkaloids from Malaysian Mitragyna speciosa (Kratom) and the role of oxindoles on human opioid receptors', *Journal of Natural Products*, 84(4), pp. 1034–1043. Available at: https://doi.org/10.1021/acs.jnatprod.0c01055.

Juanda, E., Andayani, S. and Maftuch, M. (2019) 'Phytochemical screening and antibacterial activity of Kratom Leaf (Mitragyna speciosa Korth.) against Aeromonas hydrophilla', *The Journal of Experimental Life Sciences*, 9(3), pp. 155–158. Available at: https://doi.org/10.21776/UB.JELS.2019.009.03.02.

Kang, T.H. et al. (2002) 'Pteropodine and isopteropodine positively modulate the function of rat muscarinic M1 and 5-HT2 receptors expressed in Xenopus oocyte', *European Journal of Pharmacology*, 444(1–2), pp. 39–45. Available at: https://doi.org/10.1016/S0014-2999(02)01608-4.

Kashiwada, Y. et al. (2005) 'Anti-HIV benzylisoquinoline alkaloids and flavonoids from the leaves of Nelumbo nucifera, and structure-activity correlations with related alkaloids', *Bioorganic and Medicinal Chemistry*, 13(2). Available at: https://doi.org/10.1016/j.bmc.2004.10.020.

Kashyap, P. et al. (2020) 'Ajmalicine and reserpine: Indole alkaloids as multi-target directed ligands towards factors implicated in Alzheimer's disease', *Molecules*, 25(7). Available at: https://doi.org/10.3390/molecules25071609.

Kawakami, Z. et al. (2011) 'Yokukansan, a kampo medicine, protects against glutamate cytotoxicity due to oxidative stress in PC12 cells', *Journal of Ethnopharmacology*, 134(1). Available at: https://doi.org/10.1016/j.jep.2010.11.063.

Kronstrand, R. et al. (2011) 'Unintentional fatal intoxications with mitragynine and o-desmethyltramadol from the herbal blend krypton', *Journal of Analytical Toxicology*, 35(4). Available at: https://doi.org/10.1093/anatox/35.4.242.

Kruegel, A.C. et al. (2016) 'Synthetic and receptor signaling explorations of the Mitragyna alkaloids: Mitragynine as an atypical molecular framework for opioid receptor modulators', *Journal of the American Chemical Society*, 138(21). Available at: https://doi.org/10.1021/jacs.6b00360.

Kumarnsit, E., Keawpradub, N. and Nuankaew, W. (2007) 'Effect of Mitragyna speciosa aqueous extract on ethanol withdrawal symptoms in mice', *Fitoterapia*, 78(3), pp. 182–185. Available at: https://doi.org/10.1016/J.FITOTE.2006.11.012.

Kurnik-Łucka, M. et al. (2020) 'Salsolinol—neurotoxic or neuroprotective?', *Neurotoxicity Research*, 37(2). Available at: https://doi.org/10.1007/s12640-019-00118-7.

Lagunin, A.A. et al. (2020) 'Combined network pharmacology and virtual reverse pharmacology approaches for identification of potential targets to treat vascular dementia', *Scientific Reports*, 10(1). Available at: https://doi.org/10.1038/s41598-019-57199-9.

Lai, N.J.Y. et al. (2022) 'Carrageenophyte Kappaphycus malesianus inhibits microglia-mediated neuroinflammation via suppression of AKT/NF-κB and ERK signaling pathways', *Marine Drugs*, 20(8). Available at: https://doi.org/10.3390/md20080534.

La-Up, A., Saengow, U. and Aramrattana, A. (2021) 'High serum high-density lipoprotein and low serum triglycerides in Kratom users: A study of Kratom users in Thailand'. Available at: https://doi.org/10.1016/j.heliyon.2021.e06931.

Lee, H. et al. (2017) 'Isorhynchophylline, a potent plant alkaloid, induces apoptotic and anti-metastatic effects in human hepatocellular carcinoma cells through the modulation of diverse cell signaling cascades', *International Journal of Molecular Sciences*, 18(5). Available at: https://doi.org/10.3390/ijms18051095.

Li, H. et al. (2023) 'Protective effect of L-pipecolic acid on constipation in C57BL/6 mice based on gut microbiome and serum metabolomic', *BMC Microbiology*, 23(1). Available at: https://doi.org/10.1186/s12866-023-02880-3.

Li, H.Q. et al. (2019) 'Isorhynchophylline ameliorates cognitive impairment via modulating amyloid pathology, tau hyperphosphorylation and neuroinflammation: Studies in a transgenic mouse model of Alzheimer's disease', *Brain, Behavior, and Immunity*, 82. Available at: https://doi.org/10.1016/j.bbi.2019.08.194.

Li, N.M. et al. (2019) 'Mutations of beta-amyloid precursor protein alter the consequence of Alzheimer's disease pathogenesis', *Neural Regeneration Research*, 14(4). Available at: https://doi.org/10.4103/1673-5374.247469.

Lü, Q. et al. (2009) 'Guattegaumerine protects primary cultured cortical neurons against oxidative stress injury induced by hydrogen peroxide concomitant with serum deprivation', *Cellular and Molecular Neurobiology*, 29(3). Available at: https://doi.org/10.1007/s10571-008-9327-5.

Martinez-Peinado, N. et al. (2022) 'Anti-Trypanosoma cruzi activity of alkaloids isolated from Habranthus brachyandrus (Amaryllidaceae) from Argentina', *Phytomedicine*, 101. Available at: https://doi.org/10.1016/j.phymed.2022.154126.

Mat, N.H. et al. (2023) 'Analgesic effects of main indole alkaloid of kratom, mitragynine in acute pain animal model', *Behavioural Brain Research*, 439. Available at: https://doi.org/10.1016/j.bbr.2022.114251.

Matsumoto, K. et al. (1996) 'Antinociceptive action of mitragynine in mice: Evidence for the involvement of supraspinal opioid receptors', *Life Sciences*, 59(14). Available at: https://doi.org/10.1016/0024-3205(96)00432-8.

Matsumoto, K. et al. (2005) 'Inhibitory effect of mitragynine, an analgesic alkaloid from Thai herbal medicine, on neurogenic contraction of the vas deferens', *Life Sciences*, 78(2). Available at: https://doi.org/10.1016/j.lfs.2005.04.042.

Matsumoto, K., Hatori, Y., et al. (2006) 'Involvement of μ-opioid receptors in antinociception and inhibition of gastrointestinal transit induced by 7-hydroxymitragynine, isolated from Thai herbal medicine Mitragyna speciosa', *European Journal of Pharmacology*, 549(1–3). Available at: https://doi.org/10.1016/j.ejphar.2006.08.013.

Matsumoto, K., Takayama, H., et al. (2006) 'Partial agonistic effect of 9-hydroxycorynantheidine on μ-opioid receptor in the guinea-pig ileum', *Life Sciences*, 78(19). Available at: https://doi.org/10.1016/j.lfs.2005.09.030.

McGaraughty, S. et al. (2009) 'Antagonism of TRPV1 receptors indirectly modulates activity of thermoregulatory neurons in the medial preoptic area of rats', *Brain Research*, 1268. Available at: https://doi.org/10.1016/j.brainres.2009.02.018.

Meizlish, M.L. et al. (2021) 'Tissue homeostasis and inflammation', *Annual Review of Immunology*. Available at: https://doi.org/10.1146/annurev-immunol-061020-053734.

Meng, J. et al. (2023) 'Effects of hirsuteine on MDA-MB-453 breast cancer cell proliferation', *Oncology Letters*, 25(1). Available at: https://doi.org/10.3892/ol.2022.13590.

Ngernsaengsaruay, C. et al. (2022) 'Additional knowledge on the genus Mitragyna (Rubiaceae) in Thailand', *Thai Forest Bulletin (Botany)*, 50. Available at: https://doi.org/10.20531/tfb.2022.50.1.03.

Obeng, S. et al. (2020) 'Investigation of the adrenergic and opioid binding affinities, metabolic stability, plasma protein binding properties, and functional effects of selected indole-based Kratom alkaloids', *Journal of Medicinal Chemistry*, 63(1). Available at: https://doi.org/10.1021/acs.jmedchem.9b01465.

Oguntibeju, O.O. (2019) 'Type 2 diabetes mellitus, oxidative stress and inflammation: Examining the links', *International Journal of Physiology, Pathophysiology and Pharmacology*, 11(3).

Oliva, A. et al. (2003) 'Natural fungicides from Ruta graveolens L. leaves, including a new quinolone alkaloid', *Journal of Agricultural and Food Chemistry*, 51(4). Available at: https://doi.org/10.1021/jf0259361.

Osorio-Cruz, Y. et al. (2024) 'The autoxidized mixture of (-)-Epicatechin contains procyanidins and shows antiproliferative and apoptotic activity in breast cancer cells', *Pharmaceuticals (Basel, Switzerland)*, 17(2). Available at: https://doi.org/10.3390/PH17020258.

Pacifici, G. (2015) 'Clinical pharmacology of theophylline in preterm infants: Effects, metabolism and pharmacokinetics', *Current Pediatric Reviews*, 10(4). Available at: https://doi.org/10.2174/1573396311666150113213352.

Park, E.J. et al. (2014) 'Inhibitory effect of a callophycin A derivative on iNOS expression via inhibition of Akt in lipopolysaccharide-stimulated RAW 264.7 cells', *Journal of Natural Products*, 77(3). Available at: https://doi.org/10.1021/np400800h.

Parthasarathy, S. et al. (2009) 'Evaluation of antioxidant and antibacterial activities of aqueous, methanolic and alkaloid extracts from Mitragyna speciosa (rubiaceae family) leaves', *Molecules*, 14(10). Available at: https://doi.org/10.3390/molecules14103964.

Phongprueksapattana, S. et al. (2008) 'Mitragyna speciosa: Hairy root culture for triterpenoid production and high yield of mitragynine by regenerated plants', *Zeitschrift fur Naturforschung - Section C Journal of Biosciences*, 63(9–10). Available at: https://doi.org/10.1515/znc-2008-9-1014.

Pilarski, R. et al. (2007) 'Antiproliferative activity of various Uncaria tomentosa preparations on HL-60 promyelocytic leukemia cells', *Pharmacological Reports*, 59(5).

Pipatrattanaseree, W. et al. (2023) 'Exploring the enzymatic-based biological activities of Kratom leaves extracts on alpha-amylase, alpha-glucosidase, angiotensin-converting enzyme, and cholinesterases', *Journal of Thai Traditional & Alternative Medicine*, 21(3). Available at: https://he01.tci-thaijo.org/index.php/JTTAM/article/view/262319/180644 (Accessed: 22 January 2024).

Postema, P.G. et al. (2010) 'Local depolarization abnormalities are the dominant pathophysiologic mechanism for Type 1 electrocardiogram in Brugada Syndrome. A study of electrocardiograms, vectorcardiograms, and body surface potential maps during Ajmaline provocation', *Journal of the American College of Cardiology*, 55(8). Available at: https://doi.org/10.1016/j.jacc.2009.11.033.

Qi, W. et al. (2015) 'Isolation and identification of twelve metabolites of isocorynoxeine in rat urine and their neuroprotective activities in HT22 cell assay', *Planta Medica*, 81(1). Available at: https://doi.org/10.1055/s-0034-1383357.

Qing, Z.-X. et al. (2017) 'Isoquinoline alkaloids and their antiviral, antibacterial, and antifungal activities and structure-activity relationship', *Current Organic Chemistry*, 21(18). Available at: https://doi.org/10.2174/1385272821666170207114214.

Rocca, C.J. et al. (2015) 'BRCA2 is needed for both repair and cell cycle arrest in mammalian cells exposed to S23906, an anticancer monofunctional DNA binder', *Cell Cycle*, 14(13). Available at: https://doi.org/10.1080/15384101.2015.1042632.

Sabetghadam, A., Ramanathan, S. and Mansor, S.M. (2010) 'The evaluation of antinociceptive activity of alkaloid, methanolic, and aqueous extracts of Malaysian Mitragyna speciosa Korth leaves in rats', *Pharmacognosy Research*, 2(3). Available at: https://doi.org/10.4103/0974-8490.65514.

Saidin, N.A. et al. (2015) 'The cellular toxicology of mitragynine, the dominant alkaloid of the narcotic-like herb, Mitragyna speciosa Korth', *Toxicology Research*, 4(5). Available at: https://doi.org/10.1039/c5tx00113g.

Saingam, D. et al. (2013) 'Pattern and consequences of krathom (Mitragyna speciosa Korth.) use among male villagers in southern Thailand: A qualitative study', *International Journal of Drug Policy*, 24(4). Available at: https://doi.org/10.1016/j.drugpo.2012.09.004.

Salim, H.M. et al. (2021) 'Antibacterial mechanism of Kratom (Mitragyna speciosa) methanol extract on Streptococcus pneumoniae and Eschericia coli bacteria', *Biomolecular and Health Science Journal*, 4(2). Available at: https://doi.org/10.20473/bhsj.v4i2.28933.

Sengnon, N. et al. (2023) 'Seasonal and geographic variation in alkaloid content of Kratom (Mitragyna speciosa (Korth.) Havil.) from Thailand', *Plants*, 12(4). Available at: https://doi.org/10.3390/plants12040949.

Sergent, J.A. et al. (2022) 'Effects of two natural bisbenzylisoquinolines, curine and guattegaumerine, extracted from Isolona hexaloba on rhodamine efflux by Abcb1b from rat glycocholic-acid-resistant hepatocarcinoma cells', *Molecules*, 27(9). Available at: https://doi.org/10.3390/molecules27093030.

Sethi, R. et al. (2020) 'Kratom (Mitragyna speciosa): Friend or foe?', Primary Care Companion for CNS Disorders. Available at: https://doi.org/10.4088/PCC.19nr02507.

Shaik Mossadeq, W.M. et al. (2009) 'Anti-inflammatory and antinociceptive effects of Mitragyna speciosa Korth methanolic extract', *Medical Principles and Practice*, 18(5). Available at: https://doi.org/10.1159/000226292.

Shang, Y. et al. (2015) 'Activation of κ-opioid receptor exerts the glucose-homeostatic effect in streptozotocin-induced diabetic mice', *Journal of Cellular Biochemistry*, 116(2). Available at: https://doi.org/10.1002/jcb.24962.

Shay, J. et al. (2015) 'Molecular mechanisms and therapeutic effects of (-)-epicatechin and other polyphenols in cancer, inflammation, diabetes, and neurodegeneration', *Oxidative Medicine and Cellular Longevity*. Available at: https://doi.org/10.1155/2015/181260.

Simunkova, M. et al. (2019) 'Management of oxidative stress and other pathologies in Alzheimer's disease', Archives of Toxicology. Available at: https://doi.org/10.1007/s00204-019-02538-y.

Singh, D., Müller, C.P., et al. (2018) 'Evaluating the hematological and clinical-chemistry parameters of kratom (Mitragyna speciosa) users in Malaysia', *Journal of Ethnopharmacology*, 214. Available at: https://doi.org/10.1016/j.jep.2017.12.017.

Singh, D., Abdullah, M.F.I.L., et al. (2018) 'Substance use disorder related to Kratom (Mitragyna speciosa) use in Malaysia', *Current Psychopharmacology*, 8(1). Available at: https://doi.org/10.2174/2405461503666180420120649.

Singh, D. et al. (2019) 'Long-term cognitive effects of Kratom (Mitragyna speciosa Korth.) use', *Journal of Psychoactive Drugs*, 51(1). Available at: https://doi.org/10.1080/02791072.2018.1555345.

Singh, D. et al. (2022) 'The use of Kratom (Mitragyna speciosa Korth.) among people who co-use heroin and methamphetamine in Malaysia', *Journal of Addiction Medicine*, 16(2). Available at: https://doi.org/10.1097/ADM.0000000000000876.

Singh, D., Müller, C.P. and Vicknasingam, B.K. (2014) 'Kratom (Mitragyna speciosa) dependence, withdrawal symptoms and craving in regular users', *Drug and Alcohol Dependence*, 139. Available at: https://doi.org/10.1016/j.drugalcdep.2014.03.017.

Sornsenee, P., Chimplee, S. and Romyasamit, C. (2023) 'Evaluation of antibacterial, antibiofilm, antioxidant, and anti-inflammatory activities of Kratom leaves (Mitragyna speciosa) fermentation supernatant containing Lactobacillus rhamnosus GG', *Probiotics and Antimicrobial Proteins*, 1, p. 3. Available at: https://doi.org/10.1007/s12602-023-10142-x.

Suhaimi, F.W. et al. (2023) 'Kratom (*M. speciosa*) exposure during adolescence caused long-lasting cognitive behavioural deficits associated with perturbated brain metabolism pathways in adult rats', *Behavioural Brain Research*, 446. Available at: https://doi.org/10.1016/j.bbr.2023.114411.

Sun, Q. et al. (2024) 'Development of a novel UHPLC-MS/MS method for the quantification of corynoxeine: Application to pharmacokinetics and tissue distribution studies in normal and chronic unpredictable mild stress-induced depression rats', *Journal of Pharmaceutical and Biomedical Analysis*, 238, p. 115850. Available at: https://doi.org/10.1016/J.JPBA.2023.115850.

Sun, Z., Ma, C. and Zhan, X. (2023) 'Ajmalicine induces the pyroptosis of hepatoma cells to exert the antitumor effect'. Available at: https://doi.org/10.1002/jbt.23614.

Suwanlert, S. (1975) 'A study of kratom eaters in Thailand', *Bulletin on Narcotics*, 27(3).

Takayama, H. (2004) 'Chemistry and pharmacology of analgesic indole alkaloids from the rubiaceous plant, Mitragyna speciosa', *Chemical and Pharmaceutical Bulletin*. Available at: https://doi.org/10.1248/cpb.52.916.

Tan, M.A. and An, S.S.A. (2020) 'Neuroprotective potential of the oxindole alkaloids isomitraphylline and mitraphylline in human neuroblastoma SH-SY5Y cells', *3 Biotech*, 10(12). Available at: https://doi.org/10.1007/s13205-020-02535-4.

Todd, D.A. et al. (2020) 'Chemical composition and biological effects of kratom (Mitragyna speciosa): In vitro studies with implications for efficacy and drug interactions', *Scientific Reports*, 10(1). Available at: https://doi.org/10.1038/s41598-020-76119-w.

Trevisan, G. et al. (2009) 'Antinociceptive effects of 14-membered cyclopeptide alkaloids', *Journal of Natural Products*, 72(4). Available at: https://doi.org/10.1021/np800377y.

Tsuchiya, S. et al. (2002) 'Effect of mitragynine, derived from Thai folk medicine, on gastric acid secretion through opioid receptor in anesthetized rats', *European Journal of Pharmacology*, 443(1–3). Available at: https://doi.org/10.1016/S0014-2999(02)01588-1.

Tuntiyasawasdikul, S. et al. (2024) 'Mitragyna speciosa ethanolic extract: Extraction, anti-inflammatory, cytotoxicity, and transdermal delivery assessments', *Industrial Crops and Products*, 208, p. 117909. Available at: https://doi.org/10.1016/J.INDCROP.2023.117909.

Utar, Z. et al. (2011) 'Mitragynine inhibits the COX-2 mRNA expression and prostaglandin E 2 production induced by lipopolysaccharide in RAW264.7 macrophage cells', *Journal of Ethnopharmacology*, 136(1). Available at: https://doi.org/10.1016/j.jep.2011.04.011.

Váradi, A. et al. (2016) 'Mitragynine/corynantheidine pseudoindoxyls as opioid analgesics with Mu agonism and delta antagonism, which do not recruit β-Arrestin-2', *Journal of Medicinal Chemistry*, 59(18). Available at: https://doi.org/10.1021/acs.jmedchem.6b00748.

Veeramohan, R. et al. (2023) 'Comparative metabolomics analysis reveals alkaloid repertoires in young and mature Mitragyna speciosa (Korth.) Havil. leaves', *PLoS ONE*, 18(3 March). Available at: https://doi.org/10.1371/journal.pone.0283147.

Vicknasingam, B. et al. (2020) 'Kratom and pain tolerance: A randomized, placebo-controlled, double-blind study', *Yale Journal of Biology and Medicine*, 93(2).

Wang, C. et al. (2022) 'Isorhynchophylline ameliorates stress-induced emotional disorder and cognitive impairment with modulation of NMDA receptors', *Frontiers in Neuroscience*, 16. Available at: https://doi.org/10.3389/fnins.2022.1071068.

Wang, L. et al. (2020) 'Antioxidative and antiapoptosis: Neuroprotective effects of dauricine in Alzheimer's disease models', *Life Sciences*, 243. Available at: https://doi.org/10.1016/j.lfs.2019.117237.

Wang, S. et al. (2018) 'Bacteriostatic effect of quercetin as an antibiotic alternative in vivo and its antibacterial mechanism in vitro', *Journal of Food Protection*, 81(1). Available at: https://doi.org/10.4315/0362-028X.JFP-17-214.

Watanabe, K. et al. (1997) 'Inhibitory effect of mitragynine, an alkaloid with analgesic effect from Thai medicinal plant Mitragyna speciosa, on electrically stimulated contraction of isolated guinea-pig ileum through the opioid receptor', *Life Sciences*, 60(12). Available at: https://doi.org/10.1016/S0024-3205(97)00023-4.

Wearne, T.A. and Cornish, J.L. (2018) 'A comparison of methamphetamine-induced psychosis and schizophrenia: A review of positive, negative, and cognitive symptomatology', *Frontiers in Psychiatry*, 9(OCT), p. 410640. Available at: https://doi.org/10.3389/FPSYT.2018.00491/BIBTEX.

Yang, B. et al. (2023) 'Kratom use disorder and unfolded protein response: Evaluating their relationship in a case control study', *PLoS ONE*, 18(6 June). Available at: https://doi.org/10.1371/journal.pone.0287466.

Yao, M. et al. (2023) 'Isoliensinine attenuates renal fibrosis and inhibits TGF-β1/Smad2/3 signaling pathway in spontaneously hypertensive rats', *Drug Design, Development and Therapy*, 17. Available at: https://doi.org/10.2147/DDDT.S414179.

Yin, W. et al. (2003) 'Analgesic and anti-inflammatory properties of brucine and brucine N-oxide extracted from seeds of Strychnos nux-vomica', *Journal of Ethnopharmacology*, 88(2–3). Available at: https://doi.org/10.1016/S0378-8741(03)00224-1.

Yuniarti, R. et al. (2020) 'Characterization, phytochemical screenings and antioxidant activity test of Kratom leaf ethanol extract (Mitragyna speciosa Korth) using DPPH method', Journal of Physics: Conference Series. Available at: https://doi.org/10.1088/1742-6596/1462/1/012026.

Yusoff, N.H.M. et al. (2016) 'Abuse potential and adverse cognitive effects of mitragynine (kratom)', *Addiction Biology*, 21(1). Available at: https://doi.org/10.1111/adb.12185.

Zacharias, D.E., Rosenstein, R.D. and Jeffrey, G.A. (1965) 'The structure of mitragynine hydroiodide', *Acta Crystallographica*, 18(6). Available at: https://doi.org/10.1107/s0365110x65002499.

Zailan, N.F.Z., Sarchio, S.N.E. and Hassan, M. (2022) 'Evaluation of phytochemical composition, antioxidant and anti-diabetic activities of Mitragyna speciosa Methanolic Extract (MSME)', *Malaysian Journal of Medicine and Health Sciences*, 18. Available at: https://doi.org/10.47836/mjmhs.18.s21.15.

Zhang, P. et al. (2023) 'Antidiabetic and antioxidant activities of Mitragyna speciosa (kratom) leaf extract in type 2 diabetic rats', *Biomedicine and Pharmacotherapy*, 162. Available at: https://doi.org/10.1016/j.biopha.2023.114689.

Zhang, S., Ren, Y. and Qiu, J. (2018) 'Dauricine inhibits viability and induces cell cycle arrest and apoptosis via inhibiting the PI3K/Akt signaling pathway in renal cell carcinoma cells', *Molecular Medicine Reports*, 17(5). Available at: https://doi.org/10.3892/mmr.2018.8732.

13 *Piper nigrum L* in Thai Traditional Medicine

Mani Iyer Prasanth, Kanika Verma, Rajasekharan Sharika, Pankaj Nair Gayathri, Infanta Jereena Leonard, Tewin Tencomnao, and Siriporn Chuchawankul

13.1 INTRODUCTION

Black pepper (*Piper nigrum* Linn) is commonly called "Black gold" or "King of spices" due to its utilisation in international trade and economy. The cultivation originates from Southeast Asian countries, including India and Thailand. It has been a part of cuisine and traditional and folk medicine for centuries due to its potential health benefits and treatment of ailments when consumed as food, raw, or in concoctions. The various pharmacological activities exhibited by black pepper include antioxidant, anti-obesity, antitumour, antipyretic, anticonvulsant, anti-thyroid, antifungal, antibacterial, insecticidal, hepatoprotective, antiasthmatic, larvicidal, antihypertensive, anti-inflammatory, antidiabetic, antidiarrheal, bioavailability enhancer, immunomodulator, antiepileptic, antifertility, G.I. stimulant, lipid metabolism accelerator, anticancer, CNS stimulant, diuretic, aphrodisiac, blood purifier, and antiplatelet activities, among others (Joshi et al., 2019). It has been used as an immune enhancer and a treatment for gastric problems, including diarrhoea and indigestion. Apart from that, it is also used against insomnia and epilepsy (Butt et al., 2013), as well as its purpose in perfumery, as preservatives, and as insecticides (Goswami & Malviya, 2020; Srinivasan, 2007). The recent trend of using phytochemical components as possible therapeutic agents to mitigate chronic and acute diseases has paved the way for understanding various constituents that can protect, prevent, and treat diseases and disorders. Several studies have also exhibited the possible therapeutic effects of black pepper and its derived bioactives. Piperine is the primary active compound identified in black pepper. Other important compounds include alkaloids, volatile oils, and oleoresins, which are known to be responsible for the pungent smell and flavour produced by it (Butt et al., 2013).

DOI: 10.1201/9781003508724-13

13.2 BOTANICAL DESCRIPTION AND DISTRIBUTION

Black pepper, a perennial climbing vine in the Piperaceae family, is widely used in cooking and medicine. The morphology of the plant includes a strong, woody stem that can grow up to 10 metres in length (Figure 13.1A). It uses aerial roots to attach itself to other plants to climb efficiently. *P. nigrum* has simple, sequentially arranged, oval leaves with a glossy, dark green surface. These leaves, which usually range in length from 5 to 10 cm, give a unique appearance to the plant.

Tiny, dense inflorescences reproduce black pepper called spikes that hold many small flowers. These spikes eventually grow into spherical drupe clusters. The drupes are green at first, but they ripen to red. These drupes become sharp-flavoured black peppercorns when dried (Figure 13.1B). The pericarp of black peppercorns contains the alkaloid piperine, primarily responsible for their distinctive spicy flavour (Ravindran, 2012; Parthasarathy et al., 2008). Additionally, the plant has bisexual flowers that are usually white to yellowish, which help insects primarily with cross-pollination (Jagtap et al., 2006).

Worldwide, tropical climates are the primary locations for the cultivation of black pepper. Brazil, Malaysia, Thailand, Vietnam, India, and Indonesia are essential production centres (Parthasarathy et al., 2024). These areas have favourable climates ideal for *P. nigrum* growth, with high temperatures and humidity. The trade of black pepper can be traced back to 3000 B.C., when the Babylonians and Assyrians traded with the people of the Indian subcontinent.

Interestingly, black pepper, along with gold and silver, was also found to be mummified by the Pharaohs. Also, it has been documented in various literatures, such as the Vedas, the Bible, and the Quran (Milenković & Stanojević, 2021).

FIGURE 13.1 (A) *Piper nigrum* L plant. Photograph kindly provided by Mr James Phungphol. (B) Harvested and dried peppercorns. Image kindly provided by Mr Wudtipong Vongthip.

Thailand's eastern and southern regions are primarily home to black pepper plantations. Black peppers from these areas are of the finest quality and are especially famous in the province of Chanthaburi. Thai black pepper stands out for its robust and fragrant flavour, making it very popular both domestically and abroad. Modern farming methods and strict quality control measures have increased crop yields and promoted sustainable cultivation practices in Thailand's recent agricultural developments (Kasem & Thapa, 2012). By increasing Thai black pepper's output and quality, such initiatives hope to make the country more competitive internationally.

13.3 ETHNOMEDICINAL USES

Renowned as the King of Spices, black pepper has been used for ethnomedical purposes for a long time in many Asian cultures. The primary alkaloid responsible for the distinctive bitterness of the spice, piperine, is thought to have therapeutic benefits. Black pepper has long been used as a remedy for many different conditions. Since it has carminative qualities that help to release gas from the gastrointestinal tract, it is used in Ayurvedic medicine to treat conditions like indigestion, constipation, and diarrhoea (Meghwal & Goswami, 2012). Black pepper is also thought to have antibacterial, anti-inflammatory, and antioxidant qualities, which makes it a treatment for respiratory conditions like coughs and colds.

Additionally, the spice is utilised in traditional Chinese medicine to treat malaria, increase appetite, and as an analgesic (Garg, 2023). Black pepper's anti-inflammatory properties also make it a component of formulations intended to treat diseases like arthritis. Numerous traditional medical systems have extensively documented the spice's effectiveness in treating respiratory conditions such as asthma and bronchitis (Dailah, 2022; Rahman et al., 2022).

Black pepper has also long been a staple of traditional Thai medicine. Natively referred to as "Prik Thai," it is frequently used in Thai herbal treatments. It is commonly used as a stimulant to help digestion and reduce bloating to manage digestive issues. Black pepper is also used in Thai traditional medicine to treat respiratory ailments, such as bronchitis and coughs, by facilitating the expulsion of mucus (Abdulazeez et al., 2016). In addition, black pepper is added to herbal compresses and balms used in Thai massage therapy to reduce muscle inflammation and pain. According to Srinivasan (2014), the powerful antioxidants of the spice are utilised in a range of herbal preparations designed to improve overall health and facilitate detoxification. In addition, the spice is also included in diet plans to boost metabolism and aid in weight management (Gregersen et al., 2013).

13.4 PHYTOCHEMISTRY

Black pepper seeds have been used as a spice in different cuisines. The characteristics of black pepper, including colour, odour, taste, and activity, are related to the phytochemical composition of black pepper seeds. The presence and proportion of starch, fatty acids, fibres, essential oils, and other compounds are regulated by the harvesting, extraction, and chemical analytical processes. Phytochemical composition in black pepper seeds extracted with ethanol, hexane, and methanol reported the abundance of monoterpenoids, sesquiterpenoids, alkaloids, phenols, and phytosterol in dried black pepper seed extracts (Butt et al., 2013).

Piperine, the most well-known bioactive compound in black pepper seeds, varied from 1% to 63% between conditions detected by GC-MS analysis (Vieira et al., 2022; Yoon et al., 2015; Zahin et al., 2021; Vadivel et al., 2018). The compound discovered in the Piperaceae family displays antioxidation, anti-inflammation, antitumour, and anti-neurodegenerative properties (Tiwari et al., 2020). The other alkaloid derivatives, including piperanine, piperettine, piperylin A, piperolein B, and pipericine, were found in small amounts (Gorgani et al., 2017).

The sesquiterpenes, including caryophyllene and copaene, which belong to the terpenes commonly in essential oils and are responsible for the aroma, were also discovered from the seeds (Jeleń & Gracka, 2015). Additionally, the essential oils had β-Caryophyllene in the range of 18.64–31.70%, followed by monoterpene compounds, including sabinene, limonene, and β-pinene (Bagheri et al., 2014). In addition, β-caryophyllene content was used as the standard compound for quality control of Viphyllin™ black pepper seed extract, which should not be lower than 30% between batches (Sudeep et al., 2021). Furthermore, piptigrine, an insecticidal amide, was elucidated from petroleum ether extracted from dried seeds at 4.3 mg/ 5 kg dry weight (Siddiqui et al., 2004). Also, polysaccharides isolated from water extract yielded 32.4 g/ 500 g of dry weight (Chun et al., 2002). According to the literature, the abundance of phytochemicals is based on black pepper sources and extraction methods, and piperine and β-caryophyllene were identified as the main phytochemical constituents along with others listed in Table 13.1.

13.5 BIOACTIVITY

Piperine and other phytochemicals isolated from black pepper have several important pharmacological and health-enhancing properties against various diseases and metabolic conditions, explained in the following sections and tabulated in Table 13.2.

13.5.1 Anticancer Activity

The increasing resistance and associated adverse effects of chemotherapeutic drugs and the recurrence of cancer have increased the interest in the scientific community to explore new agents from medicinal plants. Researchers have reported the anticancer properties of *P. nigrum* and its bioactive compounds against different cancer cell lines. The root's chloroform and petroleum ether extracts showed a cytotoxic effect against human myeloid leukaemia HL-60 cells (IC_{50} ¼ 9.8 and 11.2 μg/ mL, respectively, after 72 hours of exposure). The methanol extract of the fruit was more effective than the dichloromethane extract against the breast cancer MCF-7 and MDA-MB-231 cell lines, while the latter showed higher cytotoxicity against MDA-MB-468 cells (Sriwiriyajan et al., 2014). The ethanolic extract of the fruit was also found to exhibit an anticancer effect on three colorectal cancer cell lines (HT-29,

TABLE 13.1
The Main Phytochemical Constituents Reported in Black Pepper Seeds

Sources	Phytochemicals	Amount	Reference
Ethanol extract	Caryophyllene	2.87%	Yousuf et al., 2022
	N-hexadecanoic acid	1.77%	
	Piperine	1.37%	
	Copaene	1.22%	
	Delta-cadinene	1.22%	
	Piperine	17.65 to 35.30 mg/g of dry weight	Vieira et al., 2022
	Piperine	133.8 ± 17.6 mg/g of extract	Yoon et al., 2015
Essential oil	β-Caryophyllene	18.64 to 25.38%	Bagheri et al., 2014
	Limonene	14.95 to 15.64%	
	β-Pinene	7.27 to 9.71%	
	β-caryophyllene	31.70%	Vieira et al., 2022
	Sabinene	23.13%	
	Limonene	21.06%	
Hexane extract	Piperine	63%	Nag et al., 2020
Methanol extract	Piperine	46.10%	Zahin et al., 2021
	Copaene	12.25%	
	Piperine	49.97%	Vadivel et al., 2018
Petroleum ether	Piptigrine	4.3 mg/ 5 kg of dry weight	Sudeep et al., 2021
Aqueous extract	Polysaccharides	32.4 g/ 500 g of dry weight	Siddiqui et al., 2004

HCT-116, and HCT-15) with IC50 values of 7.4, 3.4, and 1.9 lg/ mL, respectively (Prashant et al., 2017).

Two active compounds isolated from the fruit, (-)-kusunokinin and piperlonguminine, displayed cytotoxic effects against breast and colorectal cancer cells. (-)-kusunokinin showed greater cytotoxicity to MCF-7, MDA-MB-468, and SW-620 (IC_{50} ¼ 1.18, 1.62, and 2.60 µg/mL, respectively), while piperlonguminine was more efficient on MDA-MB-231 (IC50 ¼ 71.54 lg/mL). Both compounds displayed lower cytotoxicity against normal breast cell lines with IC_{50} values higher than 11 lg/mL. Besides, the compounds induced cell apoptosis and drove cells towards the G2/M phase. Also, both compounds decreased topoisomerase II and Bcl-2. The increase in the level of p53 further increased p21, bax, cytochrome C, and caspase-8, -7, and -3 activities (Sriwiriyajan et al., 2017).

Another compound from the root, pellitorine, was found to be cytotoxic against MCF-7 (IC50 ¼ 1.8 lg/mL) and HeLa (IC50 ¼ 13 lg/mL) (Ee. et al., 2009). The *P. nigrum* compound mainly studied for its anticancer properties is piperine, which

TABLE 13.2
Medicinal Properties of Black Pepper

Sl No	Medicinal Properties	Plant Parts Used	Extraction Method	Model Used	Dosage	Biological Effect	Reference
1	Anti-cancer	Root	Chloroform and petroleum ether	Human myeloid leukaemia cells	50 μg/mL	Cytotoxic effect	Sriwiriyajan, S. et al., 2014
2	Anti-cancer	Fruit	Ethanolic	Human colorectal cancer cell	50 μg/mL to 100 μg/mL	Anticancer effect on HT-29, HCT-116, and HCT-15	Prashant, A. et al., 2017
3	Anti-cancer	Fruit	Methanol	Human breast and colorectal cancer cells	50, 100, and 200 μg/mL	Cytotoxic effect against breast and colorectal cancer cells	Sriwiriyajan et al., 2017
4	Anti-cancer	Fruit	Methanol	Human breast cancer cells	10 to 500 μg/mL	Inhibited proliferation and caused apoptosis by activation of caspase-3 and PARP cleavage	Do, M. T. et al., 2013
5	Anti-cancer	Fruit	Methanol	Human prostate cancer cells	10 to 200 μg/mL	Dose-dependent inhibition of proliferation and induced cycle arrest	Ouyang, D. Y. et al., 2013
6	Anti-cancer	Fruit	Ethanol	Mice	50 to 200 mg/kg	95.2% reduction in tumour nodule formation, serum sialic acid level	Selvendiran, K., et al., 2006
7	Anti-microbial activity	Fruit	Hydro distillation	Gram-positive bacteria	0.5 to 10 mg/mL	Loss of control over cell membrane permeability	O'Bryan CA et al., 2015
8	Anti-microbial	Fruit	Aqueous decoction	Human oral cavity		Antibacterial activity	Khan M et al., 2007

(Continued)

TABLE 13.2 CONTINUED
Medicinal Properties of Black Pepper

Sl No	Medicinal Properties	Plant Parts Used	Extraction Method	Model Used	Dosage	Biological Effect	Reference
9	Anti-microbial	Leaf and stem assisted silver nano-particles	Ethanolic extraction	Agricultural plant pathogens	100 to 1000 μg/mL	Nanoparticles showed excellent antibacterial activity against plant pathogens	Paulkumar K et al., 2014
10	Anti-microbial	Fruit	Ethanol	Human pathogenic bacteria	250 μg/mL	Presence of alkaloids, tannins, flavonoids, cardiac and cardiac glycosides shows antibacterial properties	Ganesh P et al., 2014
11	Anti-inflammatory	Fruit	Methanol	Carrageenan-induced paw edema using plethysmometer	5, 10, and 15 mg/kg	The compound showed maximum activity at a dose of 15 mg/kg after 120 min	Tasleem, F. et al., 2014
12	Anti-inflammatory	Essential oil	Methanol	Carrageenan-induced mice	500 mg/kg	Reduced paw edema by 72% in mice	Jeena, K. et al., 2014
13	Anti-inflammatory	Fruit	Methanol	Dextran-induced acute inflammation model	100, 500, and 1000 mg/kg BW	Reduced the paw thickness by 33.3, 53.3, and 73.4%, respectively	Jeena, K. et al., 2014
14	Anti-inflammatory	Essential oil	Methanol	Chronic inflammation induced by formalin	500 mg/kg	50% inhibition of paw edema at 500 mg/kg	Heerasing Takooree et al., 2019

(Continued)

TABLE 13.2 CONTINUED
Medicinal Properties of Black Pepper

Sl No	Medicinal Properties	Plant Parts Used	Extraction Method	Model Used	Dosage	Biological Effect	Reference
15	Hepatoprotective activity	Piperine	Ethanol	Mice	100 to 500 mg/kg BW	Inhibited the increased level of serum GPT and GOT	Damanhouri, Z.A. et al., 2014
16	Hepatoprotective activity	fruit	Methano	Wistar rats	100 and 200 mg/kg	Significant liver protection	Damanhouri, Z.A. et al., 2014
17	Hepatoprotective activity	Piperine	Methanol	Wistar rats	50 mg/kg body weigh	Lipid peroxidations were significantly decreased after pre-treatment	NirwaneA M et al., 2012
18	Cardiovascular	Leaves	Hydro-ethanolic extraction	Wistar rats	E250 and E500 mg/kg	Prevent weight gain and offer cardioprotective and antioxidant benefits	Mballa et al., 2021
19	Cardiovascular	Seeds	Methanol	Swiss-albino mice	2000 mg/kg	Preventing cardiovascular events such as heart attacks and strokes	Emon et al., 2020
20	Neuroprotective	Fruits	Methanol	Beta-amyloid (1–42) rat model	100 mg/kg	Prevention or treatment of Alzheimer's disease conditions	Hritcu et al., 2015
21	Neuroprotective	Fruits	Dichloromethane	pMCAO rats	100 and 200 mg/kg	Prevented ischemia-induced cellular damage	Hua S et al., 2019
22	Neuroprotective	Fruits	Dichloromethane	pMCAO rats	100 and 150 mg/kg	Reduction in cerebral ischemic injury and protected the brain	Zhang Y et al.,2020

(Continued)

TABLE 13.2 CONTINUED
Medicinal Properties of Black Pepper

Sl No	Medicinal Properties	Plant Parts Used	Extraction Method	Model Used	Dosage	Biological Effect	Reference
23	Neuroprotective	Fruits	Cold-pressed essential oil	Sprague-Dawley male rats	100 mg/kg	Antioxidant and neuroprotective effects	Mostafa et al., 2021
24	Neuroprotective	Fruits	Hexane and ethyl acetate	Human neuroblastoma cells	12.5 mg/kg/day	Reduced oxidative stress, restored MMP, and demonstrated anti-glycation and anti-Aβ fibrilisation activities	Sharma H et al., 2023
25	Neuroprotective	Fruits	Methanol	Albino rats	25 mg/kg b.w. for 10 days	Increased serotonin levels in the brain	Balakrishnan R et al., 2023
26	Hypertension	Leaves	Ethanol	Atherogenic diet-fed hamsters	1 and 0.25 g/kg	Significant increase in lipid levels, antioxidant activity	Agbor et al., 2012
27	Hypertension	Fruits	Ethanol	Wistar rats	20 mg/kg/day	Decrease in blood pressure	Hlavackova et al., 2011
28	Hypertension	Fruits	Ethanol	Hypertensive rats	50 mg/kg/day	Antihypertensive effect	Booranasubkajorn S et al., 2017
29	Depression	Fruits	Hydrodistillation	Adult male albino Swiss mice	5,10, and 50 mg/kg	Similar anxiolytic and antidepressant effects	Ghosh S et al., 2021

(*Continued*)

TABLE 13.2 CONTINUED
Medicinal Properties of Black Pepper

Sl No	Medicinal Properties	Plant Parts Used	Extraction Method	Model Used	Dosage	Biological Effect	Reference
30	Depression	Aerial parts	Methanol	Swiss albino mice	1,000, 2,000, and 3,000 mg/kg	Antidepressant activities	Emon NU et al., 2021
31	Depression	Leaves	Hydro distillation	Swiss albino mice	5 mg/kg	Antidepressant effects	Sohi PK et al., 2013
32	Antioxidant	Fruits	Methanol and chloroform	Cervical cancer cell line		*In vitro* antioxidant activity and cytotoxicity	Sruthi D et al.,. 2017
33	Antioxidant	Fruits	Methanol, ethanol, and ethyl acetate	*In vitro*	25 mg/mL	Antioxidant activity	Okmen G et al., 2017
34	Antioxidant	Leaves	Ethanol	*In vitro*	50, 100, 250, 500, and 1000 μg	Scavenging activity	Kavitha S et al., 2018
35	Antioxidant	Seed	Water, ethanol, chloroform, ethyl acetate, and essential oil	*In vitro*	2000 μg/mL	Antioxidant activity	Batubara I et al., 2020
36	Antioxidant	Fruits	Essential oil	*In vitro* and *in vivo*	237.556 mg/g	Antioxidant and hepatoprotective properties	Zhang C et al., 2021

was found to be a potent cytotoxic agent against HER2-overexpressing breast cancer cells (SKBR3 and MCF-7 cells). Piperine strongly inhibited proliferation and caused apoptosis by activating caspase-3 and PARP cleavage. Moreover, piperine inhibited the expression of the *HER2* gene at the transcriptional level. Piperine also blocked ERK1/2 signalling and reduced the expression of SREBP-1 and FAS. In addition, piperine actively suppressed EGF-induced MMP-9 expression via the inhibition of AP-1 and NF-κB activation by interfering with ERK1/2, p38 MAPK, and Akt signalling pathways, thereby causing a reduction in migration. Also, piperine pretreatment enhanced sensitisation to paclitaxel killing in HER2-overexpressing breast cancer cells (Do et al., 2013).

Piperine also showed anticancer properties against prostate cancer cells (LNCaP, PC-3, and DU145) (Ouyang et al., 2013). Treatment with piperine resulted in a dose-dependent inhibition of the proliferation of these cell lines. Piperine induced cycle arrest at G0/G1 and downregulated cyclin D1 and cyclin A. The compound also dose-dependently increased the level of p21Cip1 and p27Kip1 in both LNCaP and DU145 but not in PC-3 cells, which conformed with the more robust cell cycle arrest observed in the former two cell lines than the latter one. In addition, piperine-induced autophagy was proved by increased LC3B-II and LC3B puncta formation in LNCaP and PC-3 cells.

Piperine also re-sensitised P-glycoprotein, multidrug resistance protein 1 (MRP1), and breast cancer resistance protein (BCRP)–dependent multidrug-resistant cancer cells (Li et al., 2011). *In vivo* administration of piperine caused a 95.2% reduction in tumour nodule formation, serum sialic acid level, and serum gamma-glutamyl transpeptidase activity in mice. The compound also suppressed benzo(a)pyrene (B(a)p) induced lung cancer in mice by decreasing the levels of total protein and protein-bound carbohydrate components (hexose, hexosamine, and sialic acid) (Selvendiran et al., 2006).

13.5.2 Antimicrobial Activity

Karsha et al. (2010) reported the antibacterial activity of black pepper, with particular reference to its mode of action on bacteria. They found excellent inhibition of the growth of gram-positive bacteria like *Staphylococcus aureus*, followed by *Bacillus cereus* and *Streptococcus faecalis*. *Pseudomonas aeruginosa* was more susceptible among the gram-negative bacteria, followed by *Salmonella typhi* and *Escherichia coli*.

Gram-positive bacteria are more susceptible to the extracts as the antibacterial action is exhibited by losing control over cell membrane permeability (O'Bryan et al., 2015). Khan et al. (2007) evaluated the antibacterial potential of aqueous decoction of *P. nigrum* and other extracts against different bacterial isolates from the oral cavity of two hundred individual volunteers, wherein black pepper showed the most potent antibacterial activity. In another study by Paulkumar et al. (2014), *P. nigrum* leaf and stem-assisted green synthesis of silver nanoparticles were evaluated for their antibacterial activity against agricultural plant pathogens and observed that these silver nanoparticles showed excellent antibacterial activity against plant pathogens.

Ganesh et al. (2014) experimented with phytochemical analysis and the antibacterial activity of *P. nigrum* against human pathogenic bacteria. They noted that alkaloids, tannins, flavonoids, and cardiac glycosides showed antibacterial properties against *Staphylococcus aureus, Salmonella typhi, Escherichia coli,* and *Proteus sp.*

13.5.3 Anti-inflammatory Activity

Tasleem et al. (2014) evaluated the anti-inflammatory effect of *P. nigrum* and its active compound piperine based on carrageenan-induced paw oedema using a plethysmometer. Piperine exhibited inhibition of oedema at all doses, with maximum activity at a dose of 15 mg/kg after 120 minutes (reaction time ¼ 0.588 s). In another study, Jeena et al. (2014) observed that using a carrageenan-induced acute inflammation model, administration of 500 mg/kg *P. nigrum* essential oil significantly reduced paw oedema by 72% in mice in the third hour when compared to the control group. On the other hand, treatment with 100 mg/kg oil produced 66.1% inhibition in the third hour. Moreover, using a dextran-induced acute inflammation model, the oil, at 100, 500, and 1000 mg/kg, dose-dependently reduced the paw thickness by 33.3, 53.3, and 73.4%, respectively, at the third hour when compared to the control group. The essential oil also showed promising results compared to the standard diclofenac (49.3% inhibition at 10 mg/kg). In the case of chronic inflammation induced by formalin, the *P. nigrum* essential oil caused a 50% inhibition of paw oedema at 500 mg/kg. In comparison, the inhibition exhibited by the standard drug diclofenac at 10 mg/kg was 57.5% (Takooree et al., 2019).

13.5.4 Hepatoprotective Activity

Piperine inhibited the increased level of serum GPT and GOT in a dose-dependent manner in a hepatotoxicity model of mice caused by D-galactosamine. The hepatoprotective activity of methanolic extract of *P. nigrum* fruits was evaluated in ethanol-CCl_4–induced hepatic damage in Wistar rats. Ethanol-CCl_4 was used to induce hepatotoxicity in the rats. Prophylactic treatment with methanolic extract of *P. nigrum* at a dose of 100 and 200 mg/kg body weight, p.o. and pretreatment with piperine at a dose of 50 mg/kg body weight, p.o. for 15 days with ethanol-CCl4–treated rats showed significant liver protection as evidenced from the triglycerides levels, alanine transaminase, aspartate transaminase, alkaline phosphatase, bilirubin and superoxide dismutase, catalase, glutathione reductase, and lipid peroxidation levels to assess the liver functions (Damanhouri et al., 2014).

Administration of Ethanol-CCl_4 exhibited a significant boost in triglycerides, alanine transaminase, aspartate transaminase, alkaline phosphatase, and bilirubin levels. At the same time, there was a substantial decrease in the superoxide dismutase, catalase, and glutathione reductase levels, which were restored to normal levels after pretreatment with methanolic extract of *P. nigrum* and piperine. Lipid peroxidation was also significantly decreased after pretreatment with methanolic extract of *P. nigrum* and piperine at given doses. The results were similar to the reference standard, Liv52, at a 1 mL/kg dose, p.o. for 15 days. The morphological and histopathological studies of the liver also supported the biochemical parameters, thereby highlighting the hepatoprotective potential of *P. nigrum* (Nirwane et al., 2012).

13.5.5 Neuroprotective Activity

The methanolic extract of *P. nigrum* fruits in a beta-amyloid (1–42) rat model exhibited a significant increase in neurotransmitter levels, producing anxiolytic- and antidepressant-like effects, further validating its antioxidant potential. Furthermore, administering the methanolic extract may provide a helpful alternative or complementary approach for the prevention or treatment of Alzheimer's disease conditions (Hritcu et al., 2015). The dichloromethane extraction from *P. nigrum* reduced neurological deficits and markedly prevented ischemia-induced cellular damage in pMCAO rats and showed neuroprotective effects through significant improvements in body weight, neurological dysfunction, postural reflex, body sway, beam balancing, and grip strength (Hua S et al., 2019). The dichloromethane fraction of *P.* nigrum significantly reduced cerebral ischemic injury in a rat model of permanent middle cerebral artery occlusion (pMCAO), and the fraction (150 mg/kg) significantly reduced cerebral ischemic injury and protected the brain. Moreover, it inhibited apoptosis and autophagy by activating the Akt-mTOR signalling pathway in ischaemic stroke, demonstrating its potential for neuroprotection (Zhang et al., 2020).

Cold-pressed oil from black pepper fruits was investigated against scopolamine-induced memory impairment in rats, significantly altering the activity of antioxidant enzymes (catalase, superoxide dismutase) and reducing malondialdehyde (MDA) levels. Additionally, the oil reduced the level of acetylcholinesterase in the hippocampus by 51%, supported by histopathological and molecular studies showing that black pepper oil has antioxidant and neuroprotective potential (Mostafa et al., 2021). The extracts of various parts of *P. nigrum* showed neuroprotective effects *in vitro* (Yu et al., 2022).

The hexane and ethyl acetate extracts of black pepper in human neuroblastoma cells (SH-SY5Y) significantly reduced oxidative stress, restored mitochondrial membrane potential, and demonstrated potent anti-glycation and anti-Aβ fibrillisation activities. Additionally, they acted as competitive inhibitors of acetylcholinesterase (AChE). The multitarget neuroprotective mechanisms of *P. nigrum* suggest it is a promising candidate for treating neurodegenerative diseases (Sharma H et al., 2023). Piperine effectively prevents symptoms of neurodegenerative diseases and alterations in the health-mediated signalling pathways of the host by modulating cell survival and death and thereby aiding in the treatment of neurological disorders (Balakrishnan et al., 2023).

13.5.6 Hypertension

Black pepper is among the 60 medicinal plants used to treat hypertension and prevent or control blood pressure in the traditional medicinal system (Rajalakshmi et al., 2016). A study by Agbor et al. (2012) estimated the protective effects of *P. nigrum* on the cardiac antioxidant status of hamsters fed an atherogenic diet, which caused a significant increase in lipid levels and altered antioxidant activity. Hamsters supplemented with different doses of *P. nigrum* (1 and 0.25 g/kg) for 12 weeks significantly prevented these adverse effects, demonstrating its protective role against atherogenic diet-induced oxidative stress in the heart. *P. nigrum* is also an essential component of traditional Siddha medicine for the treatment of hypertension. Wistar rats fed with L-NAME (40 mg/kg/day) along with piperine (20 mg/kg/day) showed that the L-NAME-induced increase in blood pressure was partially reduced by piperine (Hlavackova et al., 2011). Additionally, as the active ingredient in different herbal formulas, piperine has shown its antihypertensive effect in L-NAME-induced endothelial dysfunction in mice (Booranasubkajorn et al., 2017).

13.5.7 Depression

In a study by Ghosh et al. (2021), the anxiolytic and antidepressant-like activities of *P. nigrum* essential oil were evaluated, showing similar anxiolytic and antidepressant effects when compared to the elevated plus maze and tail suspension test in rats, suggesting the dual anxiolytic and antidepressant-like effects of the essential oils possibly via the serotonergic pathway. The methanol extract of *P. nigrum* showed dose-dependent antidepressant activities using both *in vivo* forced swimming and tail suspension techniques in mice, which highlighted the potential of antidepressant drugs (Emon et al., 2021).

In another study, *P. nigrum* significantly showed anti-anxiety and antidepressant-like activities in the mouse model for depression and normalised the hypothalamic-pituitary-adrenal axis by reducing plasma corticosterone levels and modulating related genes in the frontal cortex and hippocampus (Chotritthirong et al., 2023). According to traditional Unani medicine, fruits of *P. nigrum* have pharmacologically shown antidepressant activities (Jan et al., 2024). *P. nigrum* (5 mg/kg) along with imipramine (20 mg/kg) in Swiss mice for four weeks exhibited better antidepressant effects than individual treatments and was effective and safe for antidepressant treatment (Sohi et al., 2013).

13.5.8 Antioxidant Activity

Methanol and chloroform extracts exhibited higher antioxidant activity than hexane and water extracts of dried fruits of *P. nigrum, as* reported in in vitro studies (Sruthi et al., 2017; Okmen et al., 2017). The ethanolic leaf extract of *P. nigrum* showed maximum activity in non-enzymatic assays such as DPPH and ABTS (Kavitha S et al., 2018). The essential oil extracts of *P. nigrum* demonstrated antioxidant activity, which was more effective when compared to synthetic antioxidants (Batubara et al., 2020). It exhibited significant *in vitro* and *in vivo* antioxidant properties as determined by free radical and lipid peroxidation scavenging abilities and by reducing CCl_4-induced injury in mice (Zhang et al., 2021).

13.5.9 Cardiovascular Activity

Black pepper and piperine have shown protective effects against cardiovascular diseases (Wang et al., 2021). Hydro-ethanolic extract from *P. nigrum* leaves showed preventive effects on metabolic and cardiovascular changes in Wistar rats fed a cafeteria diet. The cafeteria diet led to a 30% weight gain, increased lipid levels by over 50%, and a 90% increase in adipose tissue weight, along with oxidative stress and organ structure alterations. Treatment with *P. nigrum* extract significantly mitigated these effects. Comparable to the reference drug, *P. nigrum* leaf extract prevents weight gain and offers cardioprotective and antioxidant benefits (Mballa et al., 2021). Methanol extracts from Piper nigrum fruits have significant advantages in preventing cardiovascular events such as heart attacks and strokes (Emon et al., 2020).

13.6 Adverse Effects

Regular consumption of black pepper in different foods may not cause any toxic effects. Instead, it may show irritant effects if consumed more than usual. The absorption, metabolism, and toxicity may vary depending on the delivery system. As a nutraceutical agent, piperine can negatively impact the body if consumed in high amounts. Different tests for hepatotoxic, genotoxic, immunotoxic, and negative effects on the reproductive system have been conducted in studies involving rats, mice, or hamsters by exposing them to acute clinical doses of piperine, which did not imply any toxic effects (Haq et al., 2021). However, there have been some gaps in understanding the effects of piperine on the male reproductive system, as some contradictory studies have been reported in animal studies (Ziegenhagen et al., 2021). In this regard, it is essential to assess the existing knowledge and analyse the gaps involved in determining the effect of piperine in regulating the male reproductive system, followed by future human intervention studies by including specific investigations into this endpoint in the study design.

13.7 CONCLUSION AND FUTURE PERSPECTIVE

Black pepper seeds possess the key to alleviating several metabolic and other ailments. This chapter portrays various studies illustrating the potential phytochemicals and underlying mechanisms mediating these effects. The chief phytochemical piperine, among others, has been discussed briefly. Regular consumption of black pepper as part of the diet can help control various simple to chronic disorders, as advocated by traditional medicine and reported by modern research. However, more in-depth studies, including animal studies and clinical trials, must be done to analyse the optimum dosage of bio-actives derived from black pepper for future therapeutic uses.

REFERENCES

Abdulazeez, M. A., Sani, I., James, B. D., & Abdullahi, A. S. (2016). Black pepper (Piper nigrum L.) oils. In Victor R. Preedy (Ed.), *Essential oils in food preservation, flavor and safety* (pp. 277–285). Academic Press.

Achaya, K. T. (1994). Indian food: A historical companion. Edition 2, Oxford University Press.

Agbor, G. A., Akinfiresoye, L., Sortino, J., Johnson, R., & Vinson, J. A. (2012). Piper species protect cardiac, hepatic and renal antioxidant status of atherogenic diet fed hamsters. *Food Chemistry*, 134(3), pp. 1354–1359.

Bagheri, H., Manap, M. Y. B. A., & Solati, Z. (2014). Antioxidant activity of Piper nigrum L. essential oil extracted by supercritical CO2 extraction and hydro-distillation. *Talanta*, 121, pp. 220–228.

Balakrishnan, R., Azam, S., Kim, I.-S., & Choi, D.-K. (2023). Neuroprotective effects of black pepper and its bioactive compounds in age-related neurological disorders. *Aging and Disease*, 14(3), p. 750.

Batubara, I., Rafi, M., & Yolanda, M. (2020). Antioxidant, antibacterial, and degradation Streptococcus mutans biofilms activities of black pepper (Piper nigrum) seed extract. AIP Conference Proceedings.

Booranasubkajorn, S., Huabprasert, S., Wattanarangsan, J., Chotitham, P., Jutasompakorn, P., Laohapand, T., . . . Tripatara, P. (2017). Vasculoprotective and vasodilatation effects of herbal formula (Sahatsatara) and piperine in spontaneously hypertensive rats. *Phytomedicine*, 24, pp. 148–156.

Butt, M. S., Pasha, I., Sultan, M. T., Randhawa, M. A., Saeed, F., & Ahmed, W. (2013). Black pepper and health claims: A comprehensive treatise. *Critical Reviews in Food Science and Nutrition*, 53(9), pp. 875–886.

Chotritthirong, Y., Chulikhit, Y., Daodee, S., Boonyarat, C., Maneenet, J., Khamphukdee, C., . . . Monthakantirat, O. (2023). Possible mechanisms for the prevention of anxiety and depressive-like behavior in a chronic mild stress mouse model by the Thai herbal medicine with Nelumbo nucifera, Centella asiatica, and Piper nigrum. *Revista Brasileira de Farmacognosia*, 33(4), pp. 756–767.

Chun, H., Shin, D. H., Hong, B. S., Dai Cho, W., Cho, H. Y., & Yang, H. C. (2002). Biochemical properties of polysaccharides from black pepper. *Biological and Pharmaceutical Bulletin*, 25(9), pp. 1203–1208.

Dailah, H. G. (2022). The ethnomedicinal evidences pertaining to traditional medicinal herbs used in the treatment of respiratory illnesses and disorders in Saudi Arabia: A review. *Saudi Journal of Biological Sciences*, 29(9), p. 103386.

Damanhouri, Z. A., & Ahmad, A. (2014). A review on therapeutic potential of Piper nigrum L. Black Pepper): The king of spices. *Medicinal & Aromatic Plants*, 3(3), p. 161.

Do, M. T., Kim, H. G., Choi, J. H., Khanal, T., Park, B. H., Tran, T. P., . . . Jeong, H. G. (2013). Antitumor efficacy of piperine in the treatment of human HER2-overexpressing breast cancer cells. *Food Chemistry*, 141(3), pp. 2591–2599.

Ee, G., Lim, C., Lim, C., Rahmani, M., Shaari, K., & Bong, C. (2009). Alkaloids from Piper sarmentosum and Piper nigrum. *Natural Product Research*, 23(15), pp. 1416–1423.

Emon, N. U., Alam, S., Rudra, S., Riya, S. R., Paul, A., Hossen, S. M., . . . Ganguly, A. (2021). Antidepressant, anxiolytic, antipyretic, and thrombolytic profiling of methanol extract of the aerial part of Piper nigrum: In vivo, in vitro, and in silico approaches. *Food Science & Nutrition*, 9(2), pp. 833–846.

Emon, N. U., Kaiser, M., Islam, M., Kabir, M. F. I., Jamir, M., Uddin, M. A. J., & Islam, M. (2020). Anxiolytic and thrombolytic investigation of methanol extract of Piper nigrum L. fruits and Sesamum indicum L. seeds. *Journal of Advanced Biotechnology and Experimental Therapeutics*, 3(3), pp. 158–164.

Ganesh, P., Kumar, R. S., & Saranraj, P. (2014). Phytochemical analysis and antibacterial activity of Pepper (Piper nigrum L.) against some human pathogens. *Central European Journal of Experimental Biology*, 3(2), pp. 36–41.

Garg, A. P. (2023). Biomedical applications of black pepper, the king of spices-A review. *Biomedical Journal of Scientific & Technical Research*, 53(1), pp. 44364–44371.

Ghosh, S., Kumar, A., Sachan, N., & Chandra, P. (2021). Anxiolytic and antidepressant-like effects of essential oil from the fruits of Piper nigrum Linn. (Black pepper) in mice: Involvement of serotonergic but not GABAergic transmission system. *Heliyon*, 7(4), e06884.

Gorgani, L., Mohammadi, M., Najafpour, G. D., & Nikzad, M. (2017). Piperine—the bioactive compound of black pepper: From isolation to medicinal formulations. *Comprehensive Reviews in Food Science and Food Safety*, 16(1), pp. 124–140.

Goswami, A., & Malviya, N. (2020). Reassessing the restorative prospectives of the king of spices black pepper. *Journal of Drug Delivery and Therapeutics*, 10(3), pp. 312–321.

Gregersen, N. T., Belza, A., Jensen, M., Ritz, C., Bitz, C., Hels, O., . . . Astrup, A. (2013). Acute effects of mustard, horseradish, black pepper and ginger on energy expenditure, appetite, ad libitum energy intake and energy balance in human subjects. *British Journal of Nutrition*, 109(3), pp. 556–563.

Haq, I. U., Imran, M., Nadeem, M., Tufail, T., Gondal, T. A., & Mubarak, M. S. (2021). Piperine: A review of its biological effects. *Phytotherapy Research*, 35(2), pp. 680–700.

Hlavačková, L., Janegová, A., Uličná, O., Janega, P., Černá, A., & Babál, P. (2011). Spice up the hypertension diet-curcumin and piperine prevent remodeling of aorta in experimental L-NAME induced hypertension. *Nutrition & Metabolism*, 8, pp. 1–10.

Hritcu, L., Noumedem, J. A., Cioanca, O., Hancianu, M., Postu, P., & Mihasan, M. (2015). Anxiolytic and antidepressant profile of the methanolic extract of Piper nigrum fruits in beta-amyloid (1–42) rat model of Alzheimer's disease. *Behavioral and Brain Functions*, 11, pp. 1–13.

Hua, S., Wang, B., Chen, R., Zhang, Y., Zhang, Y., Li, T., . . . Fu, X. (2019). Neuroprotective effect of dichloromethane extraction from Piper nigrum L. and Piper longum L. on permanent focal cerebral ischemia injury in rats. *Journal of Stroke and Cerebrovascular Diseases*, 28(3), pp. 751–760.

Jacobsen, N., & Salguero, C. P. (2014). *Thai herbal medicine: Traditional recipes for health and harmony.* Simon and Schuster.

Jagtap, S., Deokule, S., & Bhosle, S. (2006). Some unique ethnomedicinal uses of plants used by the Korku tribe of Amravati district of Maharashtra, India. *Journal of Ethnopharmacology*, 107(3), pp. 463–469.

Jan, U., Kalam, M. A., Wani, N. N., Ayoub, M., Farooq, S. F., & Yaqoob, F. (2024). Filfil Siyāh (Piper nigrum): Medicinal importance in perspective of Unani medicine and pharmacological studies. *Journal of Pharmacognosy and Phytochemistry*, 13(2), pp. 652–656.

Jeena, K., Liju, V. B., Umadevi, N., & Kuttan, R. (2014). Antioxidant, anti-inflammatory and antinociceptive properties of black pepper essential oil (Piper nigrum Linn). *Journal of Essential Oil Bearing Plants*, 17(1), pp. 1–12.

Jeleń, H. H., & Gracka, A. (2015). Analysis of black pepper volatiles by solid phase microextraction–gas chromatography: A comparison of terpenes profiles with hydrodistillation. *Journal of Chromatography A*, 1418, pp. 200–209.

Joshi, D. R., Shrestha, A. C., & Adhikari, N. (2019). A review on diversified use of the king of spices: Piper nigrum (black pepper). *International Journal of Pharmaceutical Sciences and Research*, 9(10), p. 4089, 4101.

Jotisalikorn, C. (2012). *Thai spa book: The natural Asian way to health and beauty.* Tuttle Publishing.

Karsha, P. V., & Lakshmi, O. B. (2010). Antibacterial activity of black pepper (Piper nigrum Linn.) with special reference to its mode of action on bacteria. *Indian Journal of Natural Products and Resources*, 1(2), pp. 213–215.

Kasem, S., & Thapa, G. B. (2012). Sustainable development policies and achievements in the context of the agriculture sector in Thailand. *Sustainable Development*, 20(2), pp. 98–114.

Kavitha, S., Rameshkannan, D., & Mani, D. P. (2018). Analysis of antioxidant and antidiabetic activity of Piper nigrum leaf extract by in-vitro assay. *Journal of Pharmaceutical and Biological Sciences*, 13(2), pp. 53–56.

Khan, M., & Siddiqui, M. (2007). Antimicrobial activity of Piper fruits. *Natural Product Radiance*, 6(2), pp. 111–113.

LE Mballa, D., Yadang, F. S., Tchamgoue, A. D., Mba, J. R., Tchokouaha, L. R., M. Biang, E., . . . Agbor, G. A. (2021). Cafeteria diet-induced metabolic and cardiovascular changes in rats: The role of Piper nigrum leaf extract. *Evidence-Based Complementary and Alternative Medicine*, 2021(1), p. 5585650.

Li, S., Lei, Y., Jia, Y., Li, N., Wink, M., & Ma, Y. (2011). Piperine, a piperidine alkaloid from Piper nigrum re-sensitises P-gp, MRP1 and BCRP dependent multidrug resistant cancer cells. *Phytomedicine*, 19(1), pp. 83–87.

Meghwal, M., & Goswami, T. (2012). Chemical composition, nutritional, medicinal and functional properties of black pepper: A review. 1: 172. doi: 10.4172/scientificreports. 172 Page 2 of 5 Volume 1• Issue 2• 2012 for any spice and they are combinations of many compounds; in particular for black pepper major compounds responsible for the color, odor and aroma are shown in the (Table 3) which are found out from GCMS analysis [2]. It contains lignans, alkaloids, flavonoids, aromatic compounds and amides [3]. It also contains essential oil up to, 3

Milenković, A. N., & Stanojević, L. P. (2021). Black pepper: Chemical composition and biological activities. *Advanced Technologies*, 10(2), 40–50.

Mostafa, N. M., Mostafa, A. M., Ashour, M. L., & Elhady, S. S. (2021). Neuroprotective effects of black pepper cold-pressed oil on scopolamine-induced oxidative stress and memory impairment in rats. *Antioxidants*, 10(12), p. 1993.

Nag, A., & Chowdhury, R. R. (2020a). Piperine, an alkaloid of black pepper seeds can effectively inhibit the antiviral enzymes of Dengue and Ebola viruses, an in silico molecular docking study. *Virusdisease*, 31, pp. 308–315.

Nag, A., & Chowdhury, R. R. (2020b). Piperine, an alkaloid of black pepper seeds can effectively inhibit the antiviral enzymes of Dengue and Ebola viruses, an in silico molecular docking study. *Virusdisease*, 31(3), pp. 308–315.

Nair, K. P. (2020). *The geography of black pepper (Piper nigrum)*. Springer.

Nirwane, A., & Bapat, A. (2012). Effect of methanolic extract of Piper nigrum fruits in Ethanol-CCl4 induced hepatotoxicity in Wistar rats. *Medicine, Environmental Science*, 4(3), pp. 795–802.

O'Bryan, C. A., Pendleton, S. J., Crandall, P. G., & Ricke, S. C. (2015). Potential of plant essential oils and their components in animal agriculture–in vitro studies on antibacterial mode of action. *Frontiers in Veterinary Science*, 2, p. 35.

Okmen, G., Vurkun, M., Arslan, A., & Ceylan, O. (2017). The antibacterial activities of Piper nigrum L. against mastitis pathogens and its antioxidant activities. *Indian Journal of Pharmaceutical Education and Research*, 51, pp. S170–S175.

Ouyang, D.-Y., Zeng, L.-H., Pan, H., Xu L.-h., Wang, Y., Liu, K.-P., & He, X.-H. (2013). Piperine inhibits the proliferation of human prostate cancer cells via induction of cell cycle arrest and autophagy. *Food and Chemical Toxicology*, 60, pp. 424–430.

Parthasarathy, V. A., Chempakam, B., & Zachariah, T. J. (2008). *Chemistry of spices*. Cabi.

Parthasarathy, V. A., Srinivasan, V., Mathew, P. J., Neema, V. P., Krishnamurthy, K. S., Shivakumar, M. S., ... Ravindran, P. N. (2024). Black pepper. In *Handbook of Spices in India: 75 Years of Research and Development* (pp. 1393–1576). Singapore: Springer Nature Singapore.

Paulkumar, K., Gnanajobitha, G., Vanaja, M., Rajeshkumar, S., Malarkodi, C., Pandian, K., & Annadurai, G. (2014). Piper nigrum leaf and stem assisted green synthesis of silver nanoparticles and evaluation of its antibacterial activity against agricultural plant pathogens. *The Scientific World Journal*, 2014(1), p. 829894.
Prashant, A., Rangaswamy, C., Yadav, A. K., Reddy, V., Sowmya, M., & Madhunapantula, S. (2017). In vitro anticancer activity of ethanolic extracts of Piper nigrum against colorectal carcinoma cell lines. *International Journal of Applied and Basic Medical Research*, 7(1), pp. 67–72.
Rahman, M. M., Bibi, S., Rahaman, M. S., Rahman, F., Islam, F., Khan, M. S., . . . Maeesa, S. K. (2022). Natural therapeutics and nutraceuticals for lung diseases: Traditional significance, phytochemistry, and pharmacology. *Biomedicine & Pharmacotherapy*, 150, p. 113041.
Rajalakshmi, P., Vadivel, V., & Brindha, P. (2016). Review on medicinal plants recommended in Siddha literatures for the management of hypertension. *International Journal of Research in Pharmaceutical Sciences*, 7(1), pp. 16–33.
Ravindran, P., & Kallupurackal, J. A. (2012). *Black pepper Handbook of herbs and spices* (pp. 86–115). Elsevier.
Selvendiran, K., Singh, J. P. V., & Sakthisekaran, D. (2006). In vivo effect of piperine on serum and tissue glycoprotein levels in benzo (a) pyrene induced lung carcinogenesis in Swiss albino mice. *Pulmonary Pharmacology & Therapeutics*, 19(2), pp. 107–111.
Sharma, H., Sharma, N., & An, S. S. A. (2023). Black pepper (Piper nigrum) alleviates oxidative stress, exerts potential anti-glycation and anti-AChE activity: A multitargeting neuroprotective agent against neurodegenerative diseases. *Antioxidants*, 12(5), p. 1089.
Siddiqui, B. S., Gulzar, T., Begum, S., & Afshan, F. (2004). Piptigrine, a new insecticidal amide from Piper nigrum Linn. *Natural Product Research*, 18(5), pp. 473–477.
Smitinand, T., & Santisuk, T. (1981). The spices and essential oil crops of Thailand. *Journal of the Thailand Research Society: Natural History Supplement*, 29(01685), p. 85.
Sohi, P. K., Parashar, B., & Kansal, D. (2013). Comparative evaluation of anti-depressant activity of Glycyrrhiza glabra and Piper nigrum combination with individual drugs effects and standard drug imipramine in mice. *The International Journal of Pharmaceutical Research and Allied Sciences*, 2(4), pp. 16–21.
Spence, C. (2024). The king of spices: On pepper's pungent pleasure. *International Journal of Gastronomy and Food Science*, p. 100900.
Srinivasan, K. (2007). Black pepper and its pungent principle-piperine: A review of diverse physiological effects. *Critical Reviews in Food Science and Nutrition*, 47(8), pp. 735–748.
Srinivasan, K. 2014. Antioxidant potential of spices and their active constituents. *Critical Reviews in Food Science and Nutrition*, 54(3), pp. 352–72.
Sriwiriyajan, S., Ninpesh, T., Sukpondma, Y., Nasomyon, T., & Graidist, P. (2014). Cytotoxicity screening of plants of genus Piper in breast cancer cell lines. *Tropical Journal of Pharmaceutical Research*, 13(6), pp. 921–928.
Sriwiriyajan, S., Sukpondma, Y., Srisawat, T., Madla, S., & Graidist, P. (2017). (–)-Kusunokinin and piperloguminine from Piper nigrum: An alternative option to treat breast cancer. *Biomedicine & Pharmacotherapy*, 92, pp. 732–743.
Sruthi, D., & T Zachariah, J. (2017). In vitro antioxidant activity and cytotoxicity of sequential extracts from selected black pepper (Piper nigrum L.) varieties and Piper species. *Food Research Journal*, 24(1), pp. 75–85.

Sudeep, H., Venkatakrishna, K., Amritharaj, Gouthamchandra, K., Reethi, B., Naveen, P., . . . Shyamprasad, K. (2021). A standardised black pepper seed extract containing β -caryophyllene improves cognitive function in scopolamine-induced amnesia model mice via regulation of brain-derived neurotrophic factor and MAPK proteins. *Journal of Food Biochemistry*, 45(12), p. e13994.

Takooree, H., Aumeeruddy, M. Z., Rengasamy, K. R., Venugopala, K. N., Jeewon, R., Zengin, G., & Mahomoodally, M. F. (2019). A systematic review on black pepper (Piper nigrum L.): From folk uses to pharmacological applications. *Critical Reviews in Food Science and Nutrition*, 59(supl), pp. S210–S243.

Tasleem, F., Azhar, I., Ali, S. N., Perveen, S., & Mahmood, Z. A. (2014). Analgesic and anti-inflammatory activities of Piper nigrum L. *Asian Pacific Journal of Tropical Medicine*, 7, pp. S461–S468.

Thankamani, C. (2008). Black pepper. *Organic Spices*, p. 250.

Tiwari, A., Mahadik, K. R., & Gabhe, S. Y. (2020). Piperine: A comprehensive review of methods of isolation, purification, and biological properties. *Medicine in Drug Discovery*, 7, p. 100027.

Toussaint-Samat, M. (2009). *A history of food*. John Wiley & Sons.

Vadivel, V., Ravichandran, N., Rajalakshmi, P., Brindha, P., Gopal, A., & Kumaravelu, C. (2018). Microscopic, phytochemical, HPTLC, GC–MS and NIRS methods to differentiate herbal adulterants: Pepper and papaya seeds. *Journal of Herbal Medicine*, 11, pp. 36–45.

Varghese, R., & Ray, J. (2024). Sustainability of black pepper production: A critical analysis of physicochemical soil parameters concerning variables in pepper fields of south India. *Ecological Frontiers*, 4(4), pp. 788–801.

Vieira, L. V., Juvenato, M. E. M., Krause, M., Heringer, O. A., Ribeiro, J. S., Brandão, G. P., . . . Carneiro, M. T. W. (2022). The effects of drying methods and harvest season on piperine, essential oil composition, and multi-elemental composition of black pepper. *Food Chemistry*, 390, p. 133148.

Wang, D., Zhang, L., Huang, J., Himabindu, K., Tewari, D., Horbańczuk, J. O., . . . Atanasov, A. G. (2021). Cardiovascular protective effect of black pepper (Piper nigrum L.) and its major bioactive constituent piperine. *Trends in Food Science & Technology*, 117, pp. 34–45.

Yoon, Y. C., Kim, S.-H., Kim, M. J., Yang, H. J., Rhyu, M.-R., & Park, J.-H. (2015). Piperine, a component of black pepper, decreases eugenol-induced cAMP and calcium levels in non-chemosensory 3T3-L1 cells. *FEBS Open Bio*, 5(1), pp. 20–25.

Yousuf, M., Khan, H. M. S., Rasool, F., Khan, K. U. R., Usman, F., Ghalloo, B. A., . . . Aadil, R. M. (2022). Chemical profiling, formulation development, in vitro evaluation and molecular docking of Piper nigrum Seeds extract loaded Emulgel for anti-Aging. *Molecules*, 27(18), p. 5990.

Yu, L., Hu, X., Xu, R., Ba, Y., Chen, X., Wang, X., . . . Wu, X. (2022). Amide alkaloids characterisation and neuroprotective properties of Piper nigrum L.: A comparative study with fruits, pericarp, stalks and leaves. *Food Chemistry*, 368, p. 130832.

Zahin, M., Bokhari, N. A., Ahmad, I., Husain, F. M., Althubiani, A. S., Alruways, M. W., . . . Shalawi, M. (2021). Antioxidant, antibacterial, and antimutagenic activity of Piper nigrum seeds extracts. *Saudi Journal of Biological Sciences*, 28(9), pp. 5094–5105.

Zhang, C., Zhao, J., Famous, E., Pan, S., Peng, X., & Tian, J. (2021). Antioxidant, hepatoprotective and antifungal activities of black pepper (Piper nigrum L.) essential oil. *Food Chemistry*, 346, p. 128845.

Zhang, Y., He, Q., Yang, M., Hua, S., Ma, Q., Guo, L., . . . Liu, J. (2020). Dichloromethane extraction from Piper nigrum L. and P. longum L. to mitigate ischemic stroke by activating the AKT/mTOR signaling pathway to suppress autophagy. *Brain Research*, 1749, p. 147047.
Ziegenhagen, R., Heimberg, K., Lampen, A., & Hirsch-Ernst, K. I. (2021). Safety aspects of the use of isolated piperine ingested as a bolus. *Foods*, 10(9), p. 2121.

14 Medicinal Roles and Health Benefits of *Citrus hystrix* DC. (Kaffir lime)

Anchalee Prasansuklab, Kamonwan Chaikhong, and Sawarin Chumpolphant

14.1 INTRODUCTION

Citrus hystrix de Candolle (*Citrus hystrix* DC.) is an aromatic herbal plant that belongs to the Rutaceae family. It is commonly known as Kaffir lime, Makrut (in Thailand), Limau Purut (in Malaysia), Jeruk Purut (in Indonesia), Kabuyau (in the Philippines), Trúc (in Vietnam), and Krauch soeuch (in Cambodia). The leaves and fruit of *C. hystrix* are popularly used as key ingredients in Southeast Asian cuisine, especially in Thai dishes. Aside from its culinary purposes, this plant has long been used in traditional herbal medicine for several illnesses and is also a promising source of chemical compounds with a variety of medicinal properties and health benefits, contributing to its high therapeutic value for humans. In this chapter, we provide an overview of *C. hystrix* with regard to the plant description and distribution, traditional uses, phytochemical components, and the biological and pharmacological properties of its extracts and bioactive constituents, along with their potential medicinal applications in several human diseases and health promotion.

14.2 BOTANICAL DESCRIPTION AND ETHNOMEDICINAL USES

C. hystrix DC. (Rutaceae) is a small tree or evergreen shrub that generally grows up to 3 to 6 m (9 to 19 ft) in height and 2.5 to 3 m (8 to 9 ft) in width. It has many branches from the stem with sharp thorns. The leaves are highly aromatic, smooth, glossy, dark green, broadly ovate to ovate-oblong, and distinctively double (two-segmented) shaped. The flowers are small and fragrant with 4–5 pinkish red-white petals. The fruits are round to egg-shaped, approximately 5 to 7 cm in diameter, and often have a distinct nipple-like structure near the apex. The skin is thick, highly wrinkled and bumpy, dark green to yellowish-green when ripe, and has a strong citrus scent. The seeds are ovoid-oblong, whitish, and many in each fruit (Agouillal et al., 2017). Typically, the flowering season is restricted to spring. The fruits and leaves can be harvested year-round.

 DOI: 10.1201/9781003508724-14

FIGURE 14.1 Photograph of *Citrus hystrix* DC. (Kaffir lime). (A) Leaves (kindly provided by Mr James Phunghol) and (B) fruit.

C. hystrix (Kaffir lime) (Figure 14.1) is widely distributed in tropical regions as it grows well in a warm climate and well-drained soil, along with ample sunlight and moisture (Agouillal et al., 2017). Hence, this plant has its roots in various countries of Southeast Asia, including Thailand, Indonesia, Malaysia, Vietnam, Cambodia, and the Philippines, as well as Northeastern India and Southern China (Siti et al., 2022). According to the World Flora Online (WFO) Plant List (https://wfoplantlist.org/plant-list), *C. hystrix* DC. is the accepted scientific name along with many taxonomic synonyms such as *C. auraria* Michel, *C. boholensis* (Wester) Tanaka, *C. celebica* Koord., *C. combara* Raf., *C. echinata* St.Lag., *C. hyalopulpa* Yu.Tanaka, *C. kerrii* (Swingle) Yu.Tanaka, *C. macroptera* Montrouz., *C. micrantha* Wester, *C. papuana* F.M.Bailey, *C. tuberoides* J.W.Benn., and *C. westeri* Tanaka.

In Thailand, *C. hystrix* leaves are an important ingredient in many popular Thai dishes such as Tom Yum soup, Tom Kha soup, Thai authentic curries, Thai steamed fish curry custard, Thai corn fritters, Thai fried fish cakes, and Thai traditional crispy noodles. Besides the culinary uses, *C. hystrix* also has a long history in Thai folk medicine where different parts of the plant are utilised for treating several types of illnesses (Agouillal et al., 2017; Hutadilok-Towatana et al., 2006). The leaves and peels of *C. hystrix* have been traditionally used as antitussive, expectorant, blood purifier, antihemorrhagic, carminative, anti-flatulent, stomachic, and digestive stimulant. Fruit juice and leaves help maintain healthy teeth and gums and are used as a remedy for scurvy. This plant has also been known to alleviate flu, fever, headache, stomach-ache, abdominal pain, and hypertension. Kaffir lime peel oil and fruit juice are locally used as shampoo for lice removal, anti-dandruff, and hair loss treatment (Wisetkomolmat et al., 2021). There have been recordings that the leaf and fruit parts were used by folk healers to treat gynaecological ailments including pelvic inflammatory disease (menstrual/periodic fever) and leucorrhoea (Sitthithaworn et al., 2019).

14.3 PHYTOCHEMISTRY

A number of bioactive compounds have been reported in the peels and leaves of *C. hystrix*. These compounds can be classified into various phytochemical classes including diterpene, fatty acid, furanocoumarin, monoterpenes, monoterpenoids, phytosterols, and sesquiterpenes. (Hien et al., 2020; Long et al., 2023; Pattarachotanant & Tencomnao, 2020; Quyen et al., 2020; Sreepian et al., 2019; Sri Tunjung et al., 2018). A summary of major bioactive compounds found in *C. hystrix* is presented in Table 14.1 with their names, chemical structures, and potential medicinal uses.

Apart from major components mentioned above, there are small amounts of various other phytoconstituents identified in *C. hystrix* such as 1,2,3-Propanetriol (Glycerin), 1,7-Nonadiene, 4,8-dimethyl, 13-Methyl-tetradec-13-ene-1,12-dio, Farnesol, 17-Pentatriocontane, 1-Methyl cyclohexene, 1-Octadecene, 4-methyl-1,4-heptadiene, Benzenepropanoic acid, Borane, Cetene, cis-Linalool oxide, Cyclohexane, Cyclohexanol, Dimethylmethyl, Dotriacontane, Dotriacontanol, Eicosane, Elemol, Hydroxyfuranocoumarin, Lupeyl acetate, Monoacetate, Naphthalene, Vitamin E, β-tocopherol, Phytol acetate, 1,2-Benzenedicarboxylic acid, butyl octyl ester, Citronellyl propanoate, Methyl 6-oxoheptanoate, Hexadecanoic acid, ethyl ester, Linolenic acid, Margaric acid, 2-Thujene, Camphene, Cymene, 4-(3-Methyl-2-oxobut oxy)-7H-furo(3,2-g)(1)benzopyran-7-one (Isooxypeucedanin), 7H-Furo(3,2-g)(1) benzopyran-7-one, 4-(2,3-epoxy-3-methylbutoxy)-,(S)-(−)-, endo-Borneol, Geranyl acetate, Trans-linalool oxide, Isopulegol, α-Terpinene, α-Terpinolene, α-Thujene, β-Myrcene, β-Phellandrene, γ-Terpinene, Citronellol acetate, Cyclohexanol,2-(2-hydroxy-2-propyl)-5-methyl-, Phenol,2,4-bis(1,1-dimethylethyl)-, Campesterol, γ-Sitosterol, 1,6,10-Dodecatrien-3-ol, 3,7,11-trimethyl-, Cadina-1(10),4-diene, Delta-cadinene, D-Germacrene, Nerolidol, α-Farnesene, α-Muurolene, β-Elemene, Caryophyllene oxide, and Dihydrolanosterol (Hien et al., 2020; Long et al., 2023; Pattarachotanant & Tencomnao, 2020; Quyen et al., 2020; Sreepian et al., 2019; Sri Tunjung et al., 2018).

14.4 POTENTIAL MEDICINAL PROPERTIES AND HEALTH BENEFITS

14.4.1 Cancer

C. hystrix extract has been reported to exert anti-cancer effects in many types of cancer. The extract could induce cell cycle arrest and apoptosis in a triple-negative breast cancer cell line (MDA-MB-231) via the inhibition of anti-apoptotic Bcl-2 expression (Ho et al., 2020). The extracts were also found to inhibit the growth of breast and colorectal cancer cell lines through induction of oxidative stress and cell apoptosis (Abolmaesoomi et al., 2023) and able to reduce the viability of cervical carcinoma HeLa cells and several neuroblastoma cell lines (Tunjung et al., 2015). Several bioactive compounds found in *C. hystrix* extracts, including phytol, n-hexadecanoic acid, citronellal, citronellol, linalool, d-limonene, terpinen-4-ol, α-pinene, lupeol, and sitosterol (Geetha & Varalakshmi, 2001; Hien et al., 2020; Long et al., 2023; Pattarachotanant & Tencomnao, 2020; Quyen et al., 2020; Saleem, 2009; Sreepian

TABLE 14.1
Major Bioactive Compounds Identified in *C. hystrix*

Name	Structure	Potential Medicinal Uses
Diterpene		
Phytol		Parasitic infection (de Moraes et al., 2014), breast cancer (Boris Pejin et al., 2014), heart disease (B. Pejin et al., 2014), and diabetes mellitus (Elmazar et al., 2013)
Fatty acid		
n-Hexadecanoic acid (Palmitic acid)		Atherosclerosis (Wu et al., 2014), colon cancer (Bharath et al., 2021), and breast cancer (Sangpairoj et al., 2022)
Furanocoumarin		
4-(2,3-Dihydroxy-3-methylbutoxy) furo(3,2-g)chromen-7-one (Oxypeucedanin hydrate)	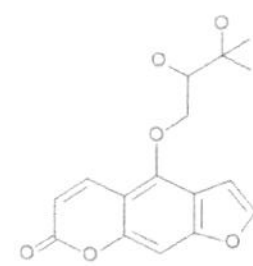	Bacterial infection and fungal infection (Dongfack et al., 2012)
Xanthotoxol	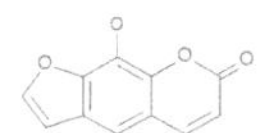	Cerebral ischaemia (He et al., 2013), neuropathic pain (Jiang et al., 2016), lung cancer (Lin et al., 2022), and skin diseases (Lee & Hyun, 2022)
Monoterpenes		
D-Limonene		Gastric cancer (Lu et al., 2004), colon cancer (Jia et al., 2013), renal damage (Rehman et al., 2014), and ulcerative colitis (Yu et al., 2017)

(Continued)

TABLE 14.1 CONTINUED
Major Bioactive Compounds Identified in *C. hystrix*

Name	Structure	Potential Medicinal Uses
Sabinene		Dermatophytosis (Valente et al., 2013), gastric ulcers (Azab et al., 2017), and skeletal muscle atrophy (Ryu et al., 2019)
Terpinen-4-ol	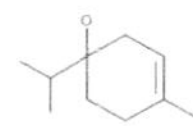	Candidiasis (Mondello et al., 2006; Ninomiya et al., 2013), lung cancer (Wu et al., 2012), gastrointestinal cancer (Shapira et al., 2016), cardiac arrhythmias (Gondim et al., 2017), and cutaneous melanoma (Di Martile et al., 2021)
α-Pinene	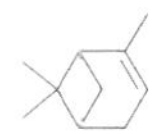	Inflammatory diseases (Kim et al., 2015), brain cancer (Aydin et al., 2013), allergic rhinitis (Nam et al., 2014), and gastric ulcers (Azab et al., 2017)
β-Pinene		Fungal infection (Feng et al., 2020) and diabetes mellitus (Santos et al., 2022)
Monoterpenoids		
Citronellal	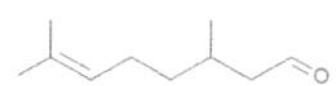	Anisakiasis (Barros et al., 2009), pain (Melo et al., 2010), inflammatory disease (Melo et al., 2011), breast cancer (Ho et al., 2020), and alzheimer's disease (K et al., 2023)
Citronellol		Inflammatory pain (Brito et al., 2012), dermatophytosis (Pereira et al., 2015), allergic diseases (Kobayashi et al., 2016), and breast cancer (Ho et al., 2020)
Citronellyl acetate	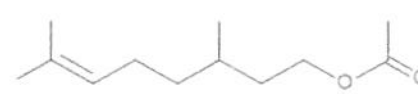	Pain (Rios et al., 2013)
Linalool	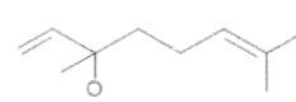	Inflammatory pain (Peana et al., 2006), carpal tunnel syndrome (Seol et al., 2016), breast cancer (Elbe et al., 2022), and candidiasis (Medeiros et al., 2022)

(Continued)

TABLE 14.1 CONTINUED
Major Bioactive Compounds Identified in *C. hystrix*

Name	Structure	Potential Medicinal Uses
α-Terpineol	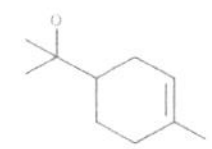	Cardiovascular diseases (Sabino et al., 2013), bacterial diseases (Srisukh et al., 2012), obesity (de Sousa et al., 2020)
Geraniol	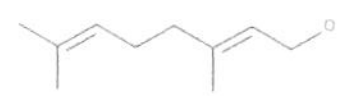	Antimicrobial activity (Lira et al., 2020; Pereira et al., 2015)
Phytosterols		
Sitosterol	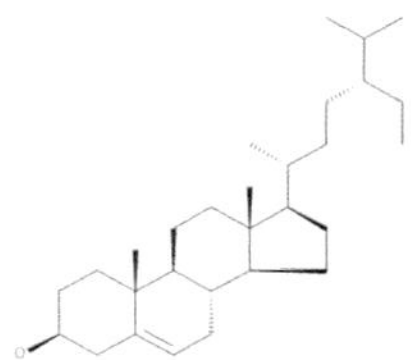	Atherosclerosis (Loizou et al., 2010), analgesic (Nirmal et al., 2012), renal cancer (Sharmila & Sindhu, 2017), and diabetes mellitus (Babu et al., 2020; Ponnulakshmi et al., 2019)
Sesquiterpenes		
Cadinene	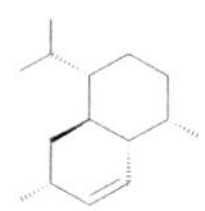	Fungal infection (Kundu et al., 2013)
β-caryophyllene	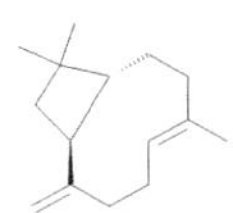	Cerebral ischaemia (Chang et al., 2013), dyslipidemic diseases (Baldissera et al., 2017), atherosclerosis (Harb et al., 2018), rheumatoid arthritis (Ames-Sibin et al., 2018), fungal infection (Nogueira Sobrinho et al., 2020), and wound healing (Gushiken et al., 2022)

(Continued)

TABLE 14.1 CONTINUED
Major Bioactive Compounds Identified in *C. hystrix*

Name	Structure	Potential Medicinal Uses
α-Humulene (α-Caryophyllene)	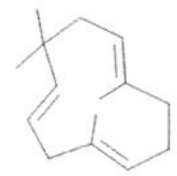	Inflammatory diseases (Fernandes et al., 2007), allergic diseases (Rogerio et al., 2009), diabetes mellitus (Gunawan et al., 2016), and hepatocellular carcinoma (Chen et al., 2019)
α-Copaene	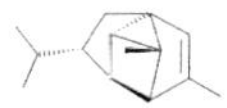	Neuroblastoma (Turkez et al., 2014)
β-Cubebene	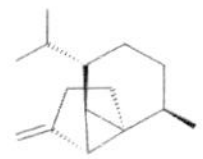	Alzheimer's disease (SevİMoĞLu et al., 2023)
Spathulenol	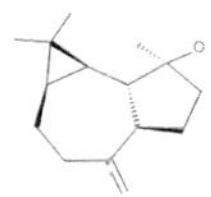	Antimycobacterial activities (Dzul-Beh et al., 2020)
Alkane		
Hexacosane	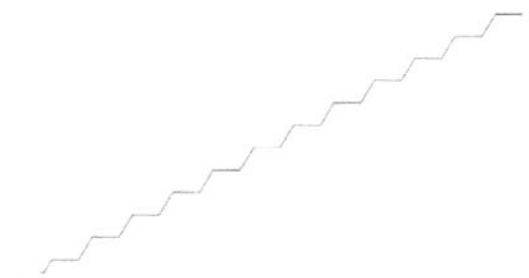	Antimicrobial activities (Rukaiyat et al., 2015)

(Continued)

TABLE 14.1 CONTINUED
Major Bioactive Compounds Identified in *C. hystrix*

Name	Structure	Potential Medicinal Uses
Coumarin		
7-Hydroxycoumarin	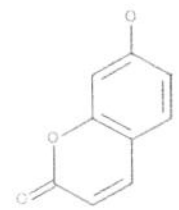	Anti-inflammatory and antioxidant effects (Sim et al., 2015)
Neophytadiene		Anti-inflammatory effects (Bhardwaj et al., 2020)
Esters		
Ethyl palmitate		Anti-inflammatory activity (Saeed et al., 2012)
Triterpene		
Lupeol	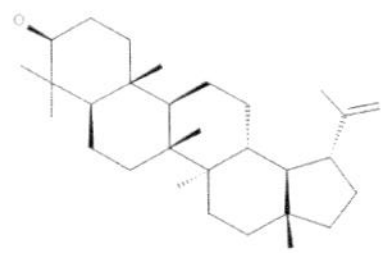	Anti-inflammatory and anti-cancer (Geetha & Varalakshmi, 2001; Saleem, 2009), atopic dermatitis (Bae et al., 2023)
Squalene	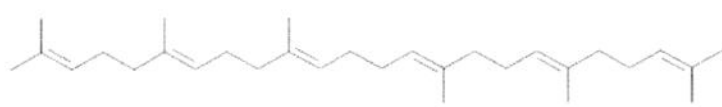	Anti-oxidant and anti-inflammation (Zhang et al., 2023)

et al., 2019; Sri Tunjung et al., 2018), have also been reported to possess anti-cancer properties in various types of cancer cells through the mechanisms of inhibiting cell viability, cell proliferation, cell cycle transition, and autophagy as well as apoptosis induction.

Phytol, a diterpene compound presented in *C. hystrix*, has been reported to decrease the viability of breast adenocarcinoma MCF-7 cell line (Boris Pejin et al., 2014). Many studies have revealed antitumour properties of other bioactive compounds in *C. hystrix*, including n-hexadecanoic acid, citronellal, citronellol, and linalool, as they could promote apoptosis, autophagy, and inhibit proliferation in human breast cancer cells (Elbe et al., 2022; Ho et al., 2020; Sangpairoj et al., 2022). Additionally, n-hexadecanoic acid has been observed to induce reactive oxygen species (ROS) generation and inhibit HT-29 colon cancer cells through the induction of apoptosis and cell cycle arrest (Bharath et al., 2021). Furthermore, d-limonene has been found to decrease Bcl-2 protein and induce apoptosis via the mitochondrial death pathway and suppress the phosphoinositide 3-kinase (PI3K)/protein kinase B (Akt) pathway in human colon carcinoma LS174T cell line (Jia et al., 2013).

The PI3K/Akt pathway is a crucial signalling pathway that is responsible for regulating essential cellular processes such as cell metabolism, cell growth, and cell apoptosis. In many cancer types, this signalling pathway becomes hyperactive, thereby promoting cancer cell survival and proliferation (Jia et al., 2013). Xanthotoxol was shown to induce cell cycle arrest and apoptosis in non-small cell lung cancer (NSCLC) cell lines through downregulation of the PI3K/Akt signalling pathway (Lin et al., 2022). Terpinen-4-ol has also demonstrated anticancer activity on NSCLC cell lines via the p53-dependent mitochondrial apoptotic pathway (Wu et al., 2012). Additionally, d-limonene, terpinen-4-ol, α-pinene, and sitosterol have shown their activities against various types of cancer cells including gastric cancer, gastrointestinal cancer, brain cancer, and renal cancer, respectively (Aydin et al., 2013; Lu et al., 2004; Shapira et al., 2016; Sharmila & Sindhu, 2017).

14.4.2 Diabetes Mellitus

Diabetes mellitus is a chronic metabolic disorder characterised by a prolonged elevated blood glucose level (hyperglycaemia), which has also been linked to multiple comorbidities such as obesity, dyslipidaemia, hypertension, cardiovascular diseases, and autoimmune diseases. Some phytocompounds contained in *C. hystrix* leaf and peel, including β-pinene, sitosterol, and phytol (Hien et al., 2020; Long et al., 2023; Pattarachotanant & Tencomnao, 2020) demonstrate antidiabetic properties that can be considered as potential therapeutic agents for the management of type-2 diabetes.

β-pinene has been reported to possess glucose-lowering and hypolipidemic effects as well as anti-inflammatory action in a diabetic rat model (Santos et al., 2022). This compound was able to lower the levels of plasma glucose, triglycerides, very-low-density lipoprotein (VLDL), low-density lipoprotein (LDL), and high-density lipoprotein (HDL). It also presented anti-inflammatory action by reducing paw edema and leukocyte migration in carrageenan-induced diabetic rats. Babu et al. (2020) showed that sitosterol could attenuate insulin resistance in the adipose tissue of high-fat diet and sucrose-fed type-2 diabetic rats by restoring glucose transporter-4 (GLUT4) expression mediated by the insulin receptor substrate 1 (IRS-1)/Akt signaling pathway (Babu et al., 2020). Meanwhile, the treatment with sitosterol was found to lower blood glucose levels and normalise the altered biochemical parameters, including serum insulin and testosterone, lipid profile, oxidative stress markers, and antioxidant enzymes (Ponnulakshmi et al., 2019). Potential antidiabetic role of phytol has been revealed by Elmazar et al. (2013) that this compound could restore glucose and lipid homeostasis, increase serum adiponectin, and decrease TNF-α levels in diabetic insulin-resistant rats (Elmazar et al., 2013).

14.4.3 Neurological Diseases

Neurological diseases are conditions of damage to the brain, spinal cord, or nerves that affect the nervous system. Hyperglycaemia is considered one of major risk factors that can trigger neuronal damage, which may later progress to encephalopathy and the development of neurodegenerative disorders through inducing neuronal senescence (Luna et al., 2021; Pattarachotanant & Tencomnao, 2020). Many studies found that *C. hystrix,* with reported antidiabetic properties, also demonstrated beneficial effects on neuroprotection.

The extracts from *C. hystrix* peel and leaves could prevent high glucose-induced senescence in the SH-SY5Y human neuroblastoma cell line. The protective mechanisms were possibly mediated by inhibiting ROS generation, inducing cell cycle progression, and upregulating sirtuin 1 (SIRT1) expression (Pattarachotanant & Tencomnao, 2020). Interestingly, the investigation of two bioactive compounds identified in *C. hystrix* peel or leaf extracts, citronellal and β-cubebene (Hien et al., 2020; Long et al., 2023; Pattarachotanant & Tencomnao, 2020), revealed both compounds as promising candidates against Alzheimer's disease or Alzheimer's dementia, the most common type of neurodegenerative disorder in older adults and characterised by a progressive loss of neuronal structure and function (Lamptey et al., 2022). Citronellal was reported to have anti-acetylcholinesterase activity through *in silico* and *in vivo* approaches. Moreover, administration of citronellal significantly improved cognitive function and the activities of several key antioxidant enzymes in a scopolamine-induced Alzheimer's mice model (K et al., 2023). In the meantime, β-cubebene has also been identified as a bioactive candidate for dementia treatment through *in silico* docking analysis against different proteins associated with dementia and Alzheimer's disease, obtained from high-throughput gene expression data sets (SevİMoĞLu et al., 2023).

Cerebral ischaemia, characterised by a reduction in blood flow and oxygen supply to the brain, is among the most common causes of neurodegenerative dementia. This pathology can occur in a particular brain region or in the whole brain, leading to neuronal injury and irreversible brain damage (Pluta, 2022). The ischaemia is often followed by a marked inflammatory response at the affected site, accompanied by the elevated expression of inflammatory mediators such as cytokines and adhesion molecules, which play an important role in the progression of brain and neuronal damage (Iadecola & Alexander, 2001). The study by He et al. (2013) found that xanthotoxol, a natural bioactive compound presented in C. hystrix, exerted neuroprotective properties after cerebral ischaemia. By using a rat model of focal cerebral ischaemia/reperfusion injury, the results showed that xanthotoxol administration could attenuate brain oedema and infarct size, blood-brain barrier disruption, and the expression of pro-inflammatory mediators such as interleukin (IL)-1β, IL-8, tumour necrosis factor (TNF)-α, nitric oxide (NO), inducible nitric oxide synthase (iNOS), cyclooxygenase-2 (COX-2), as well as the nuclear factor-κB (NF-κB) p65 subunit in the ischaemic core area of the cortex region (He et al., 2013). Similarly, β-caryophyllene was suggested to have potent neuroprotective activity against cerebral ischaemic injury by alleviating cortical infarct volume and neurological deficit, partly mediated through the suppression of inflammatory mediators including IL-1β, IL-6, COX-2, NO, and prostaglandin E2 in microglial cells (Chang et al., 2013).

14.4.4 Inflammatory Diseases

Many studies have demonstrated that *C. hystrix* and its phytoconstituents possess anti-inflammatory properties, which make this plant useful for the treatment of inflammatory diseases. The ethanolic extract from *C. hystrix* leaves, along with its isolated active compound lupeol, could attenuate inflammation in lipopolysaccharide (LPS)-stimulated macrophages by suppressing the release of IL-1β, IL-6, and TNF-α cytokines, the production of pro-inflammatory mediators, as well as the expression of NLRP3 inflammasome-associated genes (Buakaew et al., 2021). Similarly, α-pinene could significantly decrease LPS-induced production of IL-6, TNF-α, NO, iNOS, and COX-2 via the inhibition of the ERK and JNK MAPKs/NF-κB pathway in mouse peritoneal macrophage cells (Kim et al., 2015). It was observed that the anti-inflammatory effect of linalool in the LPS-stimulated macrophage cell line is mainly involved with the release of iNOS, not its expression level. The change in PGE2 release and COX2 expression also failed to be observed after linalool treatment (Peana et al., 2006).

More bioactive compounds in *C. hystrix* have demonstrated their anti-inflammatory properties in rodents. Fernandes et al. (2007) found that α-humulene displayed a marked inhibitory effect on both carrageenan-induced rat and mouse paw oedema, mediated through the regulation of expression of different inflammatory mediators including platelet-activating factor, bradykinin, histamine, IL-1β, TNFα, PGE2, COX-2, and iNOS (Fernandes et al., 2007). Citronellal and β-pinene showed anti-inflammatory potential in a rat model by reducing carrageenan-induced paw oedema and leukocyte migration to the peritoneal cavity (Melo et al., 2011; Santos et al., 2022). Brito et al. (2012) reported the anti-inflammatory activity of citronellol in a carrageenan-induced pleurisy mouse model by inhibiting neutrophil infiltration and elevated TNF-α production in pleural exudates, as well as decreasing the release of NO in LPS-stimulated murine macrophages (Brito et al., 2012).

The naturally occurring fatty acid ester named ethyl palmitate has shown promising anti-inflammatory activity in different experimental rat models in the study by Saeed et al. (2012) (Saeed et al., 2012). This was evidenced by decreased paw oedema and PGE2 levels in carrageenan-induced rat paw oedema model, diminished ear oedema and myeloperoxidase (MPO) activity in the croton oil–induced rat ear oedema model, and reduced plasma levels of TNFα and IL-6, as well as ameliorated histological changes in lung and liver tissues in the LPS-induced endotoxemia model of systemic inflammation. Neophytadiene was found to inhibit NO and pro-inflammatory cytokine production in LPS-stimulated macrophages and could also attenuate LPS-induced cardiac injury in a rat model. The expression of several inflammatory markers, including IL-6, IL-10, PGE2, IL-1β, TNF-α, and iNOS, was significantly reduced in heart tissue after treatment with neophytadiene, possibly via the PI3K/Akt- and MAPK-mediated regulation of the NF-κB signaling pathway (Bhardwaj et al., 2020).

In addition to cell lines and rodents, the study using copper sulfate–induced inflammation in zebrafish (Danio rerio) was performed to evaluate the anti-inflammatory properties of squalene (Zhang et al., 2023). This study showed that squalene could reduce the number of migratory neutrophils to the inflammatory site (forming clusters along the horizontal midline of the trunk and tail) and also downregulate the expression of genes encoding TNFα and COX-2. The study by Timonen et al. (2011) suggested the possible anti-inflammatory potential of 7-hydroxycoumarin since several synthesised 7-hydroxycoumarin derivatives were found to inhibit iNOS expression, as well as NO and IL-6 production, in LPS-induced murine macrophages (Timonen et al., 2011).

14.4.5 Atherosclerosis

Atherosclerosis is a chronic inflammatory vascular disease characterised by the build-up of cholesterol plaque inside the walls of the arteries and is considered as a primary cause of stroke, heart attack, and other cardiovascular diseases (CVD). Several pro-inflammatory mediators have been reported to play a key role in the progression of atherosclerosis, such as TNF-α, iNOS, IL-1β, IL-6, IL-22, chemokine ligand CCL2 (MCP-1), and CCL5 (RANTES), C-X-C motif chemokine ligand 12 (CXCL12), adhesion molecules, C-reactive protein (CRP), PGE2, and leukotriene B4 (LTB4). These inflammatory mediators are involved in pathological alterations of the vascular wall structure and function, called vascular remodelling, which includes endothelial dysfunction, increased vascular permeability, monocyte infiltration, foam cell generation, ultimately leading to vessel occlusion or triggering thrombosis after plaque rupture (Evans et al., 2022; Lu et al., 2011).

Some bioactive compounds present in *C. hystrix* could have potential to attenuate the development and progression of atherosclerosis. A sesquiterpene compound, β-caryophyllene, has been suggested as a protective agent in the treatment against atherosclerosis due to its cholesterol-lowering effect. The compound could reduce serum total cholesterol, LDL, and atherogenic index, while increasing HDL level in hypercholesterolaemia rat model (Harb et al., 2018). Moreover, α-Terpineol was reported to exert anti-hypertensive activity in an NG-nitro-L-arginine methyl ester (L-NAME)-induced hypertensive rat model (Sabino et al., 2013). β-sitosterol has been shown to play a role in cardiovascular protection by lowering cholesterol levels and alleviating inflammation of the endothelium (Loizou et al., 2010). Monocyte infiltration is an early process of atherosclerosis that involves the upregulation of endothelial adhesion molecules, including vascular adhesion molecule 1 (VCAM-1) and intracellular adhesion molecule 1 (ICAM-1), to facilitate the attachment of monocytes to endothelial cells, promoting their migration and contributing to plaque formation (Moriya, 2019). Loizou et al. (2010) showed that β-sitosterol could suppress the expression of both VCAM-1 and ICAM-1, as well as attenuate the phosphorylation of NF-kB p65 in TNF-α-stimulated human aortic endothelial cells (HAECs), resulting in decreased monocyte attachment to the endothelium (Loizou et al., 2010).

14.4.6 Infectious Diseases

Infectious diseases are caused by pathogenic microorganisms such as bacteria, viruses, fungi, and parasites. *C. hystrix* and its active components have been reported to exert potent inhibitory activities against several pathogens. The broad-spectrum inhibition of *C. hystrix* essential oil extracted from the peel was observed against several bacteria isolates, showing the most susceptibility to gram-positive strain *Staphylococcus aureus* and gram-negative bacteria *Elizabethkingia meningoseptica* (Long et al., 2023; Sreepian et al., 2019). The antibacterial activity of this essential oil was also demonstrated in clinical multidrug-resistant methicillin-resistant *S. aureus* (MRSA) and methicillin-susceptible *S. aureus* (MSSA) (Sreepian et al., 2023). Crude extracts from Kaffir lime leaves and callus showed the best activity against *S. aureus* and *Escherichia coli* when extracted using chloroform and ethyl acetate, respectively (Sri Tunjung et al., 2018).

The phytoconstituents reported in *C. hystrix* have also shown promising antibiotic effects against various pathogens. Hexacosane exhibited a moderately high activity level efficacy against *S. aureus*, *Proteus vulgaris*, *Salmonella typhi*, and *Klebsiella pneumoniae* (Rukaiyat et al., 2015). Geraniol presented both antifungal activity against *Trichophyton rubrum* and antibacterial activity against various species in the genus *Staphylococcus* (Lira et al., 2020; Pereira et al., 2015). In addition to antifungal (Pereira et al., 2015), citronellal was found to exhibit anthelmintic activity against parasites of the Anisakiases family (Barros et al., 2009). The furanocoumarin oxypeucedanin hydrate has been reported for its significant inhibitory activity against the bacteria Bacillus cereus and the fungi *Microsporum audouinii* and *Candida albicans* (Dongfack et al., 2012). The selective bacterial growth inhibitory and bactericidal activities of spathulenol were established on drug-resistant clinical isolates of *M. tuberculosis* compared to the reference susceptible strains (Dzul-Beh et al., 2020). Linalool has demonstrated its antifungal activity against fluconazole-resistant clinical strains of *C. albicans*, with the possible mechanism of action being its interaction with important enzymes involved in the biosynthesis and maintenance of cell wall and plasma membrane integrity, causing damage to the fungal cell structure (Medeiros et al., 2022). Terpinen-4-ol was found to be effective against both azole (fluconazole and itraconazole)-susceptible and -resistant human pathogenic *Candida* species. The ability of terpinen-4-ol to accelerate clearance from the vagina of all *Candida* strains was also observed in oophorectomised female Wistar rats with *C. albicans* vaginal infections (Mondello et al., 2006). The antifungal activity of β-caryophyllene was shown against strains of *Trichophyton* rubrum but not *C. albicans*. Interestingly, its synergistic action in the inhibition of T. rubrum was observed when combined with the standard drug ketoconazole (Nogueira Sobrinho et al., 2020). Regarding antiparasitic medication, the phytoconstituent phytol has potential as a promising drug for the treatment of Schistosomiasis, validated by its antischistosomal properties *in vitro* and in a mouse model of *Schistosomiasis mansoni* infection (de Moraes et al., 2014).

14.5 CONCLUSION

Citrus hystrix DC., also referred to as Kaffir lime, is well known for its wide range of therapeutic and health advantages that go well beyond the traditional Southeast Asian culinary usage. The plant's remarkable pharmacological properties are a result of the abundance of bioactive substances found in the aromatic parts of this plant, including leaves, fruit, and peel, which are diterpenes, monoterpenes, monoterpenoids, and sesquiterpenes. Several biological properties of this plant, such as antioxidant, anti-inflammatory, anti-cancer, and neuroprotective activities, have been linked to these compounds.

Traditional uses of Kaffir lime in treating a variety of illnesses like fever, flu, and hypertension have been confirmed by numerous scientific studies. Furthermore, this plant's ability to fight off chronic illnesses like cancer, neurological issues, inflammatory diseases, and microbial infections has been brought to light by these investigations. The reported medicinal potential of Kaffir lime highlights its importance for both conventional medicine as alternative therapy and modern medicine as a possible drug candidate in the future.

The high proportion of citronellal, citronellol, and limonene found in the essential oils derived from Kaffir lime leaves and peels makes them especially noteworthy. These phytoconstituents support the plant's wide range of biological activities as well as its potent fragrant qualities. Significant antibacterial, anti-inflammatory, and antioxidant properties have been shown for Kaffir lime essential oils, which make them suitable for use in medicinal, cosmetic, and food preservation compositions. In addition to volatile components, Kaffir lime contains non-volatile substances such as flavonoids and phenolic acids. The strong antioxidant properties of these chemicals bolster the plant's long-standing usage in the treatment of skin diseases and stomach issues. The other pharmacological actions of non-volatile substances also include lowering blood pressure, reducing pain and fever, and enhancing hair health.

Currently, the growing amount of data supports Kaffir lime's usage as a natural treatment and shows promise for the creation of novel pharmaceutical products, underscoring the significance of incorporating it into contemporary medical procedures. Kaffir lime can be considered a valuable plant that has the potential to contribute to the advancement of health and well-being through ongoing research and innovation, which will benefit a broad range of commercial and medical applications.

In summary, this chapter highlights the multifunctional roles and potential applications of Kaffir lime for health benefits through an understanding of its phytochemical components and a wide range of its medicinal properties. Future research could be focused on optimising extraction techniques to increase the yield as well as additional clinical research to investigate novel therapeutic avenues. Through these efforts, unlocking Kaffir lime's full potential can be achieved, subsequently enhancing the plant's significance as a potentially profitable crop with remarkable health advantages.

REFERENCES

Abolmaesoomi, M., Mat Junit, S., Mohd Ali, J., Chik, Z. B., & Abdul Aziz, A. (2023). Effects of polyphenolic-rich extracts from Citrus hystrix on proliferation and oxidative stress in breast and colorectal cancer. *Turkish Journal of Biochemistry*, 48(1), 110–118. https://doi.org/doi:10.1515/tjb-2022-0062

Agouillal, F., Taher, Z. M., Moghrani, H., Nasrallah, N., & El Enshasy, H. (2017). A review of genetic taxonomy, biomolecules chemistry and bioactivities of Citrus hystrix DC. *Biosciences Biotechnology Research Asia*, 14(1), 285.

Ames-Sibin, A. P., Barizão, C. L., Castro-Ghizoni, C. V., Silva, F. M. S., Sá-Nakanishi, A. B., Bracht, L., . . . Comar, J. F. (2018). β-Caryophyllene, the major constituent of copaiba oil, reduces systemic inflammation and oxidative stress in arthritic rats. *Journal of Cellular Biochemistry*, 119(12), 10262–10277. https://doi.org/https://doi.org/10.1002/jcb.27369

Aydin, E., Türkez, H., & Geyikoğlu, F. (2013). Antioxidative, anticancer and genotoxic properties of α-pinene on N2a neuroblastoma cells. *Biologia*, 68(5), 1004–1009. https://doi.org/10.2478/s11756-013-0230-2

Azab, S. S., Abdel Jaleel, G. A., & Eldahshan, O. A. (2017). Anti-inflammatory and gastroprotective potential of leaf essential oil of Cinnamomum glanduliferum in ethanol-induced rat experimental gastritis. *Pharmaceutical Biology*, 55(1), 1654–1661. https://doi.org/10.1080/13880209.2017.1314512

Babu, S., Krishnan, M., Rajagopal, P., Periyasamy, V., Veeraraghavan, V., Govindan, R., & Jayaraman, S. (2020). Beta-sitosterol attenuates insulin resistance in adipose tissue via IRS-1/Akt mediated insulin signaling in high fat diet and sucrose induced type-2 diabetic rats. *European Journal of Pharmacology*, 873, 173004. https://doi.org/https://doi.org/10.1016/j.ejphar.2020.173004

Bae, S., Jeong, N.-H., Choi, Y.-A., Lee, B., Jang, Y. H., Lee, S., & Kim, S.-H. (2023). Lupeol alleviates atopic dermatitis-like skin inflammation in 2,4-dinitrochlorobenzene/Derma tophagoides farinae extract-induced mice. *BMC Pharmacology and Toxicology*, 24(1), 27. https://doi.org/10.1186/s40360-023-00668-9

Baldissera, M. D., Souza, C. F., Grando, T. H., Doleski, P. H., Boligon, A. A., Stefani, L. M., & Monteiro, S. G. (2017). Hypolipidemic effect of β-caryophyllene to treat hyperlipidemic rats. *Naunyn-Schmiedeberg's Archives of Pharmacology*, 390(2), 215–223. https://doi.org/10.1007/s00210-016-1326-3

Barros, L., Yamanaka, A., Silva, L., Vanzeler, M., Braum, D., & Bonaldo, J. (2009). In vitro larvicidal activity of geraniol and citronellal against Contracaecum sp (Nematoda: Anisakidae). *Brazilian Journal of Medical and Biological Research*, 42, 918–920.

Bharath, B., Perinbam, K., Devanesan, S., AlSalhi, M. S., & Saravanan, M. (2021). Evaluation of the anticancer potential of Hexadecanoic acid from brown algae Turbinaria ornata on HT–29 colon cancer cells. *Journal of Molecular Structure*, 1235, 130229. https://doi.org/https://doi.org/10.1016/j.molstruc.2021.130229

Bhardwaj, M., Sali, V. K., Mani, S., & Vasanthi, H. R. (2020). Neophytadiene from Turbinaria ornata suppresses LPS-induced inflammatory response in RAW 264.7 macrophages and Sprague Dawley rats. *Inflammation*, 43, 937–950.

Brito, R. G., Guimarães, A. G., Quintans, J. S. S., Santos, M. R. V., De Sousa, D. P., Badaue-Passos, D., . . . Quintans, L. J. (2012). Citronellol, a monoterpene alcohol, reduces nociceptive and inflammatory activities in rodents. *Journal of Natural Medicines*, 66(4), 637–644. https://doi.org/10.1007/s11418-012-0632-4

Buakaew, W., Pankla Sranujit, R., Noysang, C., Thongsri, Y., Potup, P., Nuengchamnong, N., . . . Usuwanthim, K. (2021). Phytochemical constituents of Citrus hystrix DC. Leaves attenuate inflammation via NF-κB signaling and NLRP3 inflammasome activity in macrophages. *Biomolecules*, 11(1). https://doi.org/10.3390/biom11010105

Chang, H.-J., Kim, J.-M., Lee, J.-C., Kim, W.-K., & Chun, H. S. (2013). Protective effect of β-Caryophyllene, a natural bicyclic sesquiterpene, against cerebral ischemic injury. *Journal of Medicinal Food*, 16(6), 471–480. https://doi.org/10.1089/jmf.2012.2283

Chen, H., Yuan, J., Hao, J., Wen, Y., Lv, Y., Chen, L., & Yang, X. (2019). α-Humulene inhibits hepatocellular carcinoma cell proliferation and induces apoptosis through the inhibition of Akt signaling. *Food and Chemical Toxicology*, 134, 110830. https://doi.org/https://doi.org/10.1016/j.fct.2019.110830

de Moraes, J., de Oliveira, R. N., Costa, J. P., Junior, A. L. G., de Sousa, D. P., Freitas, R. M., . . . Pinto, P. L. S. (2014). Phytol, a Diterpene alcohol from chlorophyll, as a drug against neglected tropical disease Schistosomiasis Mansoni. *PLOS Neglected Tropical Diseases*, 8(1), e2617. https://doi.org/10.1371/journal.pntd.0002617

de Sousa, G. M., Cazarin, C. B. B., Junior, M. R. M., de Almeida Lamas, C., Quitete, V. H. A. C., Pastore, G. M., & Bicas, J. L. (2020). The effect of α-terpineol enantiomers on biomarkers of rats fed a high-fat diet. *Heliyon*, 6(4), e03752.

Di Martile, M., Garzoli, S., Sabatino, M., Valentini, E., D'Aguanno, S., Ragno, R., & Del Bufalo, D. (2021). Antitumor effect of Melaleuca alternifolia essential oil and its main component terpinen-4-ol in combination with target therapy in melanoma models. *Cell Death Discovery*, 7(1), 127. https://doi.org/10.1038/s41420-021-00510-3

Dongfack, M. D. J., Lallemand, M.-C., Kuete, V., Mbazoa, C., Wansi, J.-D., . . . Wandji, J. (2012). A new sphingolipid and furanocoumarins with antimicrobial activity from Ficus exasperata. *Chemical and Pharmaceutical Bulletin*, 60(8), 1072–1075. https://doi.org/10.1248/cpb.c12-00279

Dzul-Beh, A. D. J., García-Sosa, K., Uc-Cachón, A. H., Bórquez, J., Loyola, L. A., Barrios-García, H. B., . . . Molina-Salinas, G. M. (2020). In vitro growth inhibition and bactericidal activity of spathulenol against drug-resistant clinical isolates of Mycobacterium tuberculosis. *Revista Brasileira de Farmacognosia*, 29, 798–800.

Elbe, H., Ozturk, F., Yigitturk, G., Baygar, T., & Cavusoglu, T. (2022). Anticancer activity of linalool: Comparative investigation of ultrastructural changes and apoptosis in breast cancer cells. *Ultrastructural Pathology*, 46(4), 348–358. https://doi.org/10.1080/01913123.2022.2091068

Elmazar, M. M., El-Abhar, H. S., Schaalan, M. F., & Farag, N. A. (2013). Phytol/Phytanic acid and insulin resistance: Potential role of phytanic acid proven by docking simulation and modulation of biochemical alterations. *PLOS ONE*, 8(1), e45638. https://doi.org/10.1371/journal.pone.0045638

Evans, B. R., Yerly, A., van der Vorst, E. P. C., Baumgartner, I., Bernhard, S. M., Schindewolf, M., & Döring, Y. (2022). Inflammatory mediators in atherosclerotic vascular remodeling. *Frontiers in Cardiovascular Medicine*, 9, 868934. https://doi.org/10.3389/fcvm.2022.868934

Feng, X. Z., Xiao, Z., Zhang, L., Liao, S., Chen, S., Luo, H., . . . Wang, Z. (2020). Antifungal activity of β-Pinene-based hydronopyl quaternary ammonium salts against phytopathogenic fungi. *Natural Product Communications*, 15(8), 1934578X20948365. https://doi.org/10.1177/1934578x20948365

Fernandes, E. S., Passos, G. F., Medeiros, R., da Cunha, F. M., Ferreira, J., Campos, M. M., . . . Calixto, J. B. (2007). Anti-inflammatory effects of compounds alpha-humulene and (−)-trans-caryophyllene isolated from the essential oil of Cordia verbenacea. *European Journal of Pharmacology*, 569(3), 228–236. https://doi.org/https://doi.org/10.1016/j.ejphar.2007.04.059

Geetha, T., & Varalakshmi, P. (2001). Anti-inflammatory activity of lupeol and lupeol linoleate in rats. *Journal of Ethnopharmacology*, 76(1), 77–80.

Gondim, A. N. S., Lara, A., Santos-Miranda, A., Roman-Campos, D., Lauton-Santos, S., Menezes-Filho, J. E. R., . . . Cruz, J. S. (2017). (-)-Terpinen-4-ol changes intracellular Ca2+ handling and induces pacing disturbance in rat hearts. *European Journal of Pharmacology*, 807, 56–63. https://doi.org/https://doi.org/10.1016/j.ejphar.2017.04.022

Gunawan, I., Putra, A. B., & Widihati, I. (2016). The response to oxidative stress α-Humulene compounds hibiscus Manihot L leaf on the activity of 8-Hydroxy-2-Deoksiquanosin levels pancreatic β-cells in diabetic rats. *Biomedical and Pharmacology Journal*, 9(2), 433–441.

Gushiken, L. F. S., Beserra, F. P., Hussni, M. F., Gonzaga, M. T., Ribeiro, V. P., de Souza, P. F., . . . Pellizzon, C. H. (2022). Beta-caryophyllene as an antioxidant, anti-inflammatory and re-epithelialization activities in a rat skin wound excision model. *Oxidative Medicine and Cellular Longevity*, 2022, 9004014. https://doi.org/10.1155/2022/9004014

Harb, A. A., Bustanji, Y. K., & Abdalla, S. S. (2018). Hypocholesterolemic effect of β-caryophyllene in rats fed cholesterol and fat enriched diet. *Journal of Clinical Biochemistry and Nutrition*, advpub, Article 17–3. https://doi.org/10.3164/jcbn.17-3

He, W., Chen, W., Zhou, Y., Tian, Y., & Liao, F. (2013). Xanthotoxol exerts neuroprotective effects via suppression of the inflammatory response in a rat model of focal cerebral ischemia. *Cellular and Molecular Neurobiology*, 33(5), 715–722. https://doi.org/10.1007/s10571-013-9939-2

Hien, T. T., Quyen, N. T. C., Truc, T. T., & Quan, P. M. (2020). Evaluate the chemical composition of Kaffir lime (Citrus hystrix) essential oil using the classical method. *IOP Conference Series: Materials Science and Engineering*, 991(1), 012014. https://doi.org/10.1088/1757-899X/991/1/012014

Ho, Y., Suphrom, N., Daowtak, K., Potup, P., Thongsri, Y., & Usuwanthim, K. (2020). Anticancer effect of Citrus hystrix DC. Leaf extract and its bioactive constituents citronellol and, citronellal on the triple negative breast cancer MDA-MB-231 cell line. *Pharmaceuticals (Basel)*, 13(12). https://doi.org/10.3390/ph13120476

Hutadilok-Towatana, N., Chaiyamutti, P., Panthong, K., Mahabusarakam, W., & Rukachaisirikul, V. (2006). Antioxidative and free radical scavenging activities of some plants used in Thai folk medicine. *Pharmaceutical Biology*, 44(3), 221–228. https://doi.org/10.1080/13880200600685592

Iadecola, C., & Alexander, M. (2001). Cerebral ischemia and inflammation. *Current Opinion in Neurology*, 14(1), 89–94. https://doi.org/10.1097/00019052-200102000-00014

Jia, S.-S., Xi, G.-P., Zhang, M., Chen, Y.-B., Lei, B., Dong, X.-S., & Yang, Y.-M. (2013). Induction of apoptosis by D-limonene is mediated by inactivation of Akt in LS174T human colon cancer cells. *Oncology Reports*, 29(1), 349–354. https://doi.org/10.3892/or.2012.2093

Jiang, L., Wu, Q., Yang, T., & Yang, N. (2016). Xanthotoxol attenuates neuropathic pain in a rat model of chronic constriction injury. *International Journal of Clinical and Experimental Medicine*, 9(11), 21556–21561.

Kim, D. S., Lee, H. J., Jeon, Y. D., Han, Y. H., Kee, J. Y., Kim, H. J., . . . Hong, S. H. (2015). Alpha-Pinene exhibits anti-inflammatory activity through the suppression of MAPKs and the NF-κB pathway in mouse peritoneal macrophages. *American Journal of Chinese Medicine*, 43(4), 731–742. https://doi.org/10.1142/s0192415x15500457

Kobayashi, Y., Sato, H., Yorita, M., Nakayama, H., Miyazato, H., Sugimoto, K., & Jippo, T. (2016). Inhibitory effects of geranium essential oil and its major component, citronellol, on degranulation and cytokine production by mast cells. *Bioscience, Biotechnology, and Biochemistry*, 80(6), 1172–1178. https://doi.org/10.1080/09168451.2016.1148573

Kundu, A., Saha, S., Walia, S., Shakil, N. A., Kumar, J., & Annapurna, K. (2013). Cadinene sesquiterpenes from Eupatorium adenophorum and their antifungal activity. *Journal of Environmental Science and Health, Part B*, 48(6), 516–522. https://doi.org/10.1080/03601234.2013.761921

Lamptey, R. N. L., Chaulagain, B., Trivedi, R., Gothwal, A., Layek, B., & Singh, J. (2022). A review of the common neurodegenerative disorders: Current therapeutic approaches and the potential role of nanotherapeutics. *International Journal of Molecular Sciences*, 23(3). https://doi.org/10.3390/ijms23031851

Lee, Y., & Hyun, C. G. (2022). Anti-Inflammatory effects of Psoralen derivatives on RAW264.7 cells via regulation of the NF-κB and MAPK signaling pathways. *International Journal of Molecular Sciences*, 23(10). https://doi.org/10.3390/ijms23105813

Lin, X., Liu, J., Zou, Y., Tao, C., & Chen, J. (2022). Xanthotoxol suppresses non-small cell lung cancer progression and might improve patients' prognosis. *Phytomedicine*, 105, 154364. https://doi.org/https://doi.org/10.1016/j.phymed.2022.154364

Lira, M. H. P. D., Andrade Júnior, F. P. D., Moraes, G. F. Q., Macena, G. D. S., Pereira, F. D. O., & Lima, I. O. (2020). Antimicrobial activity of geraniol: An integrative review. *Journal of Essential Oil Research*, 32(3), 187–197.

Loizou, S., Lekakis, I., Chrousos, G. P., & Moutsatsou, P. (2010). β-Sitosterol exhibits anti-inflammatory activity in human aortic endothelial cells. *Molecular Nutrition & Food Research*, 54(4), 551–558. https://doi.org/https://doi.org/10.1002/mnfr.200900012

Long, D. M., Quoc, L. P. T., Nhung, T. T. P., Thy, V. B., & Nhu, N. L. Q. (2023). Chemical profiles and biological activities of essential oil of Citrus hystrix DC. peels. *Korean Journal of Food Preservation*, 30(3), 395–404. https://doi.org/10.11002/kjfp.2023.30.3.395

Lu, P. P., Liu, J. T., Liu, N., Guo, F., Ji, Y. Y., & Pang, X. (2011). Pro-inflammatory effect of fibrinogen and FDP on vascular smooth muscle cells by IL-6, TNF-α and iNOS. *Life Sciences*, 88(19–20), 839–845. https://doi.org/10.1016/j.lfs.2011.03.003

Lu, X. G., Zhan, L. B., Feng, B. A., Qu, M. Y., Yu, L. H., & Xie, J. H. (2004). Inhibition of growth and metastasis of human gastric cancer implanted in nude mice by d-limonene. *World Journal of Gastroenterology*, 10(14), 2140–2144. https://doi.org/10.3748/wjg.v10.i14.2140

Luna, R., Talanki Manjunatha, R., Bollu, B., Jhaveri, S., Avanthika, C., Reddy, N., . . . Gandhi, F. (2021). A comprehensive review of neuronal changes in diabetics. *Cureus*, 13(10), e19142. https://doi.org/10.7759/cureus.19142

Medeiros, C. I. S., Sousa, M. N. A. D., Filho, G. G. A., Freitas, F. O. R., Uchoa, D. P. L., Nobre, M. S. C., . . . Lima, E. O. (2022). Antifungal activity of linalool against fluconazole-resistant clinical strains of vulvovaginal Candida albicans and its predictive mechanism of action. *Brazilian Journal of Medical and Biological Research*, 55.

Melo, M. S., Guimarães, A. G., Santana, M. F., Siqueira, R. S., De Lima, A. D. C. B., Dias, A. S., . . . De Sousa, D. P. (2011). Anti-inflammatory and redox-protective activities of citronellal. *Biological Research*, 44(4), 363–368.

Melo, M. S., Sena, L. C. S., Barreto, F. J. N., Bonjardim, L. R., Almeida, J. R. G. S., Lima, J. T., . . . Quintans-Júnior, L. J. (2010). Antinociceptive effect of citronellal in mice. *Pharmaceutical Biology*, 48(4), 411–416. https://doi.org/10.3109/13880200903150419

Mondello, F., De Bernardis, F., Girolamo, A., Cassone, A., & Salvatore, G. (2006). In vivo activity of terpinen-4-ol, the main bioactive component of Melaleuca alternifolia Cheel (tea tree) oil against azole-susceptible and -resistant human pathogenic Candida species. *BMC Infectious Diseases*, 6, 158. https://doi.org/10.1186/1471-2334-6-158

Moriya, J. (2019). Critical roles of inflammation in atherosclerosis. *Journal of Cardiology*, 73(1), 22–27. https://doi.org/https://doi.org/10.1016/j.jjcc.2018.05.010

Nam, S.-Y., Chung, C.-K., Seo, J.-H., Rah, S.-Y., Kim, H.-M., & Jeong, H.-J. (2014). The therapeutic efficacy of α-pinene in an experimental mouse model of allergic rhinitis. *International Immunopharmacology*, 23(1), 273–282. https://doi.org/https://doi.org/10.1016/j.intimp.2014.09.010

Ninomiya, K., Hayama, K., Ishijima, S. A., Maruyama, N., Irie, H., Kurihara, J., & Abe, S. (2013). Suppression of inflammatory reactions by Terpinen-4-ol, a main constituent of tea tree oil, in a murine model of oral candidiasis and its suppressive activity to cytokine production of macrophages in vitro. *Biological and Pharmaceutical Bulletin*, 36(5), 838–844. https://doi.org/10.1248/bpb.b13-00033

Nirmal, S. A., Pal, S. C., Mandal, S. C., & Patil, A. N. (2012). Analgesic and anti-inflammatory activity of β-sitosterol isolated from Nyctanthes arbortristis leaves. *Inflammopharmacology*, 20(4), 219–224. https://doi.org/10.1007/s10787-011-0110-8

Nogueira Sobrinho, A. C., Morais, S. M. D., Souza, E. B. D., Albuquerque, M. R. J. R., Santos, H. S. D., Cavalcante, C. S. D. P., . . . Fontenelle, R. O. D. S. (2020). Antifungal and antioxidant activities of Vernonia Chalybaea Mart. ex DC. essential oil and their major constituent β-caryophyllene. *Brazilian Archives of Biology and Technology*, 63.

Pattarachotanant, N., & Tencomnao, T. (2020). Citrus hystrix extracts protect human neuronal cells against high glucose-induced senescence. *Pharmaceuticals*, 13(10), 283. https://www.mdpi.com/1424-8247/13/10/283

Pavani, K., Prasanth, D. S. N. B. K., Shadakshara, M. K. R., Ahmad, S. F., Seemaladinne, R., Rudrapal, M., & Pasala, P. K. (2023). Citronellal as a promising candidate for Alzheimer's disease treatment: A comprehensive study on in silico and in vivo anti-acetylcholine esterase activity. *Metabolites*, 13(11), 1133. https://www.mdpi.com/2218-1989/13/11/1133

Peana, A. T., Marzocco, S., Popolo, A., & Pinto, A. (2006). (–)-Linalool inhibits in vitro NO formation: Probable involvement in the antinociceptive activity of this monoterpene compound. *Life Sciences*, 78(7), 719–723. https://doi.org/https://doi.org/10.1016/j.lfs.2005.05.065

Pejin, B., Kojic, V., & Bogdanovic, G. (2014). An insight into the cytotoxic activity of phytol at in vitro conditions. *Natural Product Research*, 28(22), 2053–2056. https://doi.org/10.1080/14786419.2014.921686

Pejin, B., Savic, A., Sokovic, M., Glamoclija, J., Ciric, A., Nikolic, M., . . . Mojovic, M. (2014). Further in vitro evaluation of antiradical and antimicrobial activities of phytol. *Natural Product Research*, 28(6), 372–376. https://doi.org/10.1080/14786419.2013.869692

Pereira, F. D. O., Mendes, J. M., Lima, I. O., Mota, K. S. D. L., Oliveira, W. A. D., & Lima, E. D. O. (2015). Antifungal activity of geraniol and citronellol, two monoterpenes alcohols, against Trichophyton rubrum involves inhibition of ergosterol biosynthesis. *Pharmaceutical Biology*, 53(2), 228–234. https://doi.org/10.3109/13880209.2014.913299

Pluta, R. (2022). Ischemic brain neurodegeneration. *International Journal of Molecular Sciences*, 23(12). https://doi.org/10.3390/ijms23126441

Ponnulakshmi, R., Shyamaladevi, B., Vijayalakshmi, P., & Selvaraj, J. (2019). In silico and in vivo analysis to identify the antidiabetic activity of beta sitosterol in adipose tissue of high fat diet and sucrose induced type-2 diabetic experimental rats. *Toxicology Mechanisms and Methods*, 29(4), 276–290. https://doi.org/10.1080/15376516.2018.1545815

Quyen, N. T. C., Ngan, T. T. K., Phong, H. X., Hien, T. T., & Dung, L. T. (2020). Extraction of Kaffir Lime (Citrus hystrix DC.) essential oil by steam distillation and evaluation of chemical constituents. *IOP Conference Series: Materials Science and Engineering*, 991(1), 012015. https://doi.org/10.1088/1757-899X/991/1/012015

Rehman, M. U., Tahir, M., Khan, A. Q., Khan, R., Oday-O-Hamiza, Lateef, A., . . . Sultana, S. (2014). d-limonene suppresses doxorubicin-induced oxidative stress and inflammation via repression of COX-2, iNOS, and NFκB in kidneys of Wistar rats. *Experimental Biology and Medicine*, 239(4), 465–476. https://doi.org/10.1177/1535370213520112

Rios, E. R., Rocha, N. F., Carvalho, A. M., Vasconcelos, L. F., Dias, M. L., de Sousa, D. P., . . . Fonteles, M. M. (2013). TRP and ASIC channels mediate the antinociceptive effect of citronellyl acetate. *Chemico-Biological Interactions*, 203(3), 573–579. https://doi.org/10.1016/j.cbi.2013.03.014

Rogerio, A. P., Andrade, E. L., Leite, D. F. P., Figueiredo, C. P., & Calixto, J. B. (2009). Preventive and therapeutic anti-inflammatory properties of the sesquiterpene α-humulene in experimental airways allergic inflammation. *British Journal of Pharmacology*, 158(4), 1074–1087. https://doi.org/https://doi.org/10.1111/j.1476-5381.2009.00177.x

Rukaiyat, M., Garba, S., & Labaran, S. (2015). Antimicrobial activities of hexacosane isolated from Sanseveria liberica (Gerome and Labroy) plant. *Advancement in Medicinal Plant Research*, 3(3), 120–125.

Ryu, Y., Lee, D., Jung, S. H., Lee, K. J., Jin, H., Kim, S. J., . . . Won, K. J. (2019). Sabinene prevents skeletal muscle atrophy by inhibiting the MAPK-MuRF-1 pathway in rats. *International Journal of Molecular Sciences*, 20(19). https://doi.org/10.3390/ijms20194955

Sabino, C. K. B., Ferreira-Filho, E. S., Mendes, M. B., da Silva-Filho, J. C., Ponte, M. P. T. R., Moura, L. H. P., . . . de Oliveira, A. P. (2013). Cardiovascular effects induced by α-terpineol in hypertensive rats. *Flavour and Fragrance Journal*, 28(5), 333–339. https://doi.org/https://doi.org/10.1002/ffj.3159

Saeed, N. M., El-Demerdash, E., Abdel-Rahman, H. M., Algandaby, M. M., Al-Abbasi, F. A., & Abdel-Naim, A. B. (2012). Anti-inflammatory activity of methyl palmitate and ethyl palmitate in different experimental rat models. *Toxicology and Applied Pharmacology*, 264(1), 84–93.

Saleem, M. (2009). Lupeol, a novel anti-inflammatory and anti-cancer dietary triterpene. *Cancer Letters*, 285(2), 109–115.

Sangpairoj, K., Settacomkul, R., Siangcham, T., Meemon, K., Niamnont, N., Sornkaew, N., . . . Vivithanaporn, P. (2022). Hexadecanoic acid-enriched extract of Halymenia durvillei induces apoptotic and autophagic death of human triple-negative breast cancer cells by upregulating ER stress. *Asian Pacific Journal of Tropical Biomedicine*, 12(3), 132–140. https://doi.org/10.4103/2221-1691.338922

Santos, E. S., Abrantes Coelho, G. L., Saraiva Fontes Loula, Y. K., Saraiva Landim, B. L., Fernandes Lima, C. N., Tavares de Sousa Machado, S., . . . Kerntopf, M. R. (2022). Hypoglycemic, hypolipidemic, and anti-inflammatory effects of Beta-Pinene in diabetic rats. *Evidence-Based Complementary and Alternative Medicine*, 2022, 8173307. https://doi.org/10.1155/2022/8173307

Seol, G.-H., Kang, P., Lee, H. S., & Seol, G. H. (2016). Antioxidant activity of linalool in patients with carpal tunnel syndrome. *BMC Neurology*, 16(1), 17. https://doi.org/10.1186/s12883-016-0541-3

SevİMoĞLu, T., Atalay, V., & ÖZtÜRk, A. (2023). Determination of candidate alternative plant actives for Dementia and Alzheimer's disease proteins through Docking studies [Demans ve Alzheimer Hastalığı Proteinlerine Yönelik Aday Alternatif Bitki Aktiflerinin Kenetlenme Çalışmaları ile Belirlenmesi]. *Süleyman Demirel Üniversitesi Fen Bilimleri Enstitüsü Dergisi*, 27(2), 255–265. https://doi.org/10.19113/sdufenbed.1213583

Shapira, S., Pleban, S., Kazanov, D., Tirosh, P., & Arber, N. (2016). Terpinen-4-ol: A novel and promising therapeutic agent for human gastrointestinal cancers. *PLOS ONE*, 11(6), e0156540. https://doi.org/10.1371/journal.pone.0156540

Sharmila, R., & Sindhu, G. (2017). Evaluate the antigenotoxicity and anticancer role of β-Sitosterol by determining oxidative DNA damage and the expression of phosphorylated mitogen-activated protein kinases', C-fos, C-jun, and endothelial growth factor receptor. *Pharmacognosy Magazine*, 13(49), 95–101. https://doi.org/10.4103/0973-1296.197634

Sim, M. O., Lee, H. I., Ham, J. R., Seo, K. I., Kim, M. J., & Lee, M. K. (2015). Anti-inflammatory and antioxidant effects of umbelliferone in chronic alcohol-fed rats. *Nutrition Research and Practice*, 9(4), 364–369. https://doi.org/10.4162/nrp.2015.9.4.364

Siti, H. N., Mohamed, S., & Kamisah, Y. (2022). Potential therapeutic effects of Citrus hystrix DC and its bioactive compounds on metabolic disorders. *Pharmaceuticals (Basel)*, 15(2). https://doi.org/10.3390/ph15020167

Sitthithaworn, W., Weerasathien, L., & Onsawang, C. (2019). The use of medicinal plants for gynecologic ailments by Thai traditional folk healers in Nakhonnayok Province. *Thai Pharmaceutical and Health Science Journal*, 14(3), 111–121.

Sreepian, A., Sreepian, P. M., Chanthong, C., Mingkhwancheep, T., & Prathit, P. (2019). Antibacterial activity of essential oil extracted from Citrus hystrix (Kaffir Lime) peels: An in vitro study. *Tropical Biomedicine*, 36(2), 531–541.

Sreepian, P. M., Rattanasinganchan, P., & Sreepian, A. (2023). Antibacterial efficacy of Citrus hystrix (makrut lime) essential oil against clinical multidrug-resistant methicillin-resistant and methicillin-susceptible Staphylococcus aureus isolates. *Saudi Pharmaceutical Journal*, 31(6), 1094–1103. https://doi.org/10.1016/j.jsps.2023.03.020

Sri Tunjung, W. A., Liana, D., Hennisa, & Lisna, H. (2018). Antibacterial activity and composition of crude extracts of Kaffir Lime (Citrus hystrix DC.) leaves and Callus: Antibacterial activity of Kaffir lime leaves and Callus. *Proceedings of the Pakistan Academy of Sciences: B. Life and Environmental Sciences*, 55(2), 45–53. https://ppaspk.org/index.php/PPAS-B/article/view/183

Srisukh, V., Tribuddharat, C., Nukoolkarn, V., Bunyapraphatsara, N., Chokephaibulkit, K., Phoomniyom, S., . . . Srifuengfung, S. (2012). Antibacterial activity of essential oils from Citrus hystrix (makrut lime) against respiratory tract pathogens. *Science Asia*, 38(2), 212–217.

Timonen, J. M., Nieminen, R. M., Sareila, O., Goulas, A., Moilanen, L. J., Haukka, M., . . . Aulaskari, P. H. (2011). Synthesis and anti-inflammatory effects of a series of novel 7-hydroxycoumarin derivatives. *European Journal of Medicinal Chemistry*, 46(9), 3845–3850.

Tunjung, W. A. S., Cinatl, J., Michaelis, M., & Smales, C. M. (2015). Anti-cancer effect of Kaffir Lime (Citrus Hystrix DC) leaf extract in cervical cancer and neuroblastoma cell lines. *Procedia Chemistry*, 14, 465–468. https://doi.org/https://doi.org/10.1016/j.proche.2015.03.062

Turkez, H., Togar, B., Tatar, A., Geyıkoglu, F., & Hacımuftuoglu, A. (2014). Cytotoxic and cytogenetic effects of α-copaene on rat neuron and N2a neuroblastoma cell lines. *Biology*, 69(7), 936–942. https://doi.org/doi:10.2478/s11756-014-0393-5 (Biologia)

Valente, J., Zuzarte, M., Gonçalves, M. J., Lopes, M. C., Cavaleiro, C., Salgueiro, L., & Cruz, M. T. (2013). Antifungal, antioxidant and anti-inflammatory activities of Oenanthe crocata L. essential oil. *Food and Chemical Toxicology*, 62, 349–354. https://doi.org/https://doi.org/10.1016/j.fct.2013.08.083

Wisetkomolmat, J., Inta, A., Krongchai, C., Kittiwachana, S., Jantanasakulwong, K., Rachtanapun, P., & Rose Sommano, S. (2021). Ethnochemometric of plants traditionally utilised as local detergents in the forest dependent culture. *Saudi Journal of Biological Sciences*, 28(5), 2858–2866. https://doi.org/10.1016/j.sjbs.2021.02.018

Wu, C.-S., Chen, Y.-J., Chen, J. J. W., Shieh, J.-J., Huang, C.-H., Lin, P.-S., . . . Lin, C.-C. (2012). Terpinen-4-ol induces apoptosis in human nonsmall cell lung cancer in vitro and in vivo. *Evidence-Based Complementary and Alternative Medicine*, 2012, 818261. https://doi.org/10.1155/2012/818261

Wu, D., Liu, J., Pang, X., Wang, S., Zhao, J., Zhang, X., & Feng, L. (2014). Palmitic acid exerts pro-inflammatory effects on vascular smooth muscle cells by inducing the expression of C-reactive protein, inducible nitric oxide synthase and tumor necrosis factor-α. *International Journal of Molecular Medicine*, 34(6), 1706–1712. https://doi.org/10.3892/ijmm.2014.1942

Yu, L., Yan, J., & Sun, Z. (2017). D-limonene exhibits anti-inflammatory and antioxidant properties in an ulcerative colitis rat model via regulation of iNOS, COX-2, PGE2 and ERK signaling pathways. *Molecular Medicine Reports*, 15(4), 2339–2346. https://doi.org/10.3892/mmr.2017.6241

Zhang, P., Liu, N., Xue, M., Zhang, M., Xiao, Z., Xu, C., . . . Zhou, Y. (2023). Anti-Inflammatory and antioxidant properties of squalene in copper sulfate-induced inflammation in Zebrafish (Danio rerio). *International Journal of Molecular Sciences*, 24(10), 8518. https://www.mdpi.com/1422-0067/24/10/8518

15 Updated Medicinal Uses of Marijuana in Thailand (*Cannabis* species in Thai Traditional Medicine)

Mani Iyer Prasanth, Bhagavathi Sundaram Sivamaruthi, Rajasekharan Sharika, Infanta Jereena Leonard, Pankaj Nair Gayathri, Tewin Tencomnao, and Siriporn Chuchawankul

15.1 INTRODUCTION

In recent years, *Cannabis sativa* (Figure 15.1) has gained considerable attention for its medicinal potential, particularly in traditional medicine practices across various cultures. In Thailand, where herbal remedies are deeply rooted in the healthcare system, *C. sativa* has been explored for its therapeutic applications in treating different health conditions. This chapter delves into the botanical characteristics, cultivation, and medicinal uses of *C. sativa*, focusing on its significance in Thai traditional medicine. The plant's pharmacological properties, driven by cannabinoids and other bioactive compounds, have been studied for their efficacy in treating ailments ranging from skin disorders to more complex conditions like cancer and inflammation, underscoring its multifaceted role in historical and modern medicine.

15.2 BOTANICAL DESCRIPTION AND DISTRIBUTION

Cannabis sativa, a member of the Cannabaceae family, stands out with its four distinct subspecies: *indica*, *ruderalis*, sativa, and *afghanis*. However, the classification of these subspecies remains challenging due to the variability in their chemical and morphological characteristics, influenced by the plant's environment and pedology (Hourfane et al., 2023).

C. sativa L. is an annual, usually dioecious plant (Small and Cronquist, 1976). It is a species of the botanical genus *Cannabis* and is divided into several phenotypes (Clarke et al., 2015). *C. sativa* is a fast-growing plant with a fluted stem that can reach

 DOI: 10.1201/9781003508724-15

FIGURE 15.1 Photograph of *C. sativa* L. plant "Thai stick." Kindly provided by Mr Satit Supadee, Garden Community Enterprise integrated herb grower Band-dong-Tong. Nakhon Phanom province, Thailand.

1 to 4 m in height and a diameter ranging between 1 and 3 cm (Amaducci et al., 2008). The variation in height and diameter depends on the subspecies, environment, soil, and climatic conditions. The seeds are smooth, greyish, oval, or spherical, 2.5 to 3.5 mm long and 2.5 to 3 mm in diameter. Each seed contains two cotyledons rich in protein and oil, with albumen considered particularly small compared to other plant species (Campiglia et al., 2017; Hourfane et al., 2023).

C. sativa has long, delicate flowers. It has glandular hairs that make it fragrant and sticky (Anwar et al., 2006). At post-germination, young male and female plants cannot be distinguished. Only during the last growth phase, when flowers start appearing, does sex determination become possible (Schilling et al., 2020). The female flowers have no petals and have two long, white, yellow, or pink stigmas. Their calyx is the ovary containing a single ovule. The female flowers appear in pairs in the axils of small leaves named bracts, which contain numerous glandular trichomes where cannabinoids, mainly THC, accumulate (Lynch et al., 2016).

The male flowers have five sepals approximately 5 mm long and are yellow, white, or green. The male plants develop tiny pollen sacs that fertilise the female plants with hairy, resinous stigmas. The cannabis leaves are stipulate and opposite, with palmate, elongated, and spiny segments with toothed margins. The growth rate of the root system is relatively slow in the initial stages of vegetation, in contrast to the aerial part of the cannabis plant, which grows intensively and rapidly (Schilling et al., 2020; Strzelczyk et al., 2021; Hourfane et al., 2023).

C. sativa originates from equatorial and subtropical regions, especially central Asia (McPartland et al., 2019), where two places seem to be its cradle: the Himalayan foothills and the Pamir plains. Nonetheless, this plant has a wide geographical distribution, growing in Canada, the United States of America, Europe, and Africa. Cannabis is an ancient plant, but the craze it has generated over the last century has dramatically changed its face and even the face of the world. *C. sativa* thrives in various climates and easily adapts to different conditions. From its region of origin, it appreciates calcareous and nitrogenous soils with a neutral or slightly acidic pH. This adaptability is critical to the plant's widespread distribution (Hourfane et al., 2023).

15.3 ETHNOMEDICINAL USES

C. sativa has applications in various fields, including pharmacology, cosmetics, and nutrition (Hourfane et al., 2023). The seeds and leaves are the primary parts used for medicinal purposes. Research indicates that Cannabis seeds are edible (Callaway, 2004; Carus and Sarmento, 2016) and are used to manage nausea, vomiting, and appetite issues. Additionally, they are incorporated into cosmetics for hair strengthening (El Khomsi et al., 2022). The leaves of *C. sativa* are applied externally as poultices to treat conditions such as eczema, subcutaneous tissue disorders, and issues related to the central nervous system (Klauke et al., 2014; Mawla et al., 2012; El Khomsi et al., 2022). Despite their benefits, cannabis leaves also possess strong analgesic, sedative, and narcotic properties (Hourfane et al., 2023). The stems of *C. sativa* can be utilised as firewood and are employed in constructing tools, clothing, paper, and rope (Liu et al., 2017). Moreover, the roots address joint pain, skin burns, inflammation, vermin, and erysipelas infections (Ryz et al., 2017). They are consumed as juice to alleviate issues related to childbirth, postpartum recovery, and bleeding (Ryz et al., 2017). The leaves and inflorescences have been used as narcotics in various preparations (Knapp et al., 2019). Ethnobotanical surveys highlight the diverse uses of the entire Cannabis plant, with the aerial parts primarily addressing mental and nervous system disorders (Bouarfa et al., 2020).

15.4 PHYTOCHEMISTRY

C. sativa is rich in several bioactive compounds, and many studies have reported the secondary metabolites of cannabis and their roles. More than 100 phyto-molecules identified in the leaves, flowers, and roots of *C. sativa* belong to terpenes, and about 20 polyphenols belong to subclasses of flavone and flavonol (Flores-Sanchez and Verpoorte, 2008; Hourfane et al., 2023).

Cannabinoids are the most characterised metabolites of cannabis, despite being represented by less than 20 molecules. Cannabis seeds contain approximately 25%, 30%, and 40% proteins, fibres, and oils, respectively (Aiello et al., 2020; Taaifi et al., 2021). The oils are rich in triacylglycerols, represented by 18 different molecules, the dominating ones being LLL and OLLD, with 23% and 19%, respectively (Taaifi et al., 2021). These oils are rich in polyunsaturated fatty acids, especially linoleic, oleic, and palmitic acids (Ross et al., 1996; Babiker et al., 2021; Da Porto et al., 2012). The hydroalcoholic extract of Cannabis seeds has 12.9 ± 18.3 mg/100 g of gallic acid and 6.0 ± 5.2 mg/100 g of catechin (Babiker et al., 2021); additionally, rutin, quercetin-o-glucoside, n-trans-caffeoyltyramine, and cannabisins (a sub-class of polyphenols; cannabisins A-G, I, M-O) have been reported (Moccia et al., 2020). Cannabinoids are a group of C21 or C22 terpenolics found in cannabis, which are foremost in leaves and inflorescences. δ-5-avenasterol has also been reported in Cannabis oil. Significant sterols (lutein and β-carotene) are reported in the ethanolic extract of the seed (Irakli et al., 2019).

15.5 BIOACTIVITY

Cannabis has several important pharmacological and health-enhancing properties against various diseases and metabolic conditions, which are explained in the following sections and tabulated in Table 15.1.

15.5.1 Anticancer Activity

The active compounds found in *C. sativa*, particularly Δ9-tetrahydrocannabinol (Δ9-THC) and cannabidiol (CBD), showed that CBD can inhibit the progression of various cancers and, when combined with Δ9-THC and radiation therapy, can enhance autophagy and apoptosis in cancer cells. Additionally, cannabidiolic acid (CBDA), the acidic precursor of CBD, has been found to inhibit breast cancer cell migration (Pellati et al., 2018). A study demonstrated the impact of gamma radiation on the

TABLE 15.1
Medicinal Properties of *C. sativa*

Medicinal Property	Plant Part Used	Extraction Method	Model Used	Dosage Used	Effect Observed	Reference
Anti-cancer activity	Flower	Ethyl acetate	Cancer cells		Anti-cancer properties	Kovalchuk et al., 2020
Anti-cancer activity	Flower	Soxhlet extraction	SHSY-5Y	1, 10, 30, and 100 μg/mL	Antitumour effects	Sánchez et al., 2023
Anti-cancer activity	Leaves	Aqueous extract	MOLT-3, TALL-104, J.RT3-T3.5, and clone E6-1	329, 381, 275, and 218 μg/mL	Anti-acute leukaemia properties	Yali et al., 2021
Antimicrobial activity	Leaves	Ethanol and petroleum ether	Gram-positive and gram-negative bacteria		Strong inhibitory effects	Głodowska et al., 2017
Antimicrobial activity	Leaves	Ethanolic extract	MRSA		Inhibited MRSA growth	Chakraborty et al., 2018
Antimicrobial activity	Leaves	Acetone extraction	Bacterial culture	160 μg/mL	Showed higher zone of inhibition	Lone and Lone, 2012
Antimicrobial activity	Leaves	Aqueous extract	Gram-negative and Gram-positive bacteria	1 g /10 mL	Antibacterial activity against several human pathogens	Csakvari et al., 2021
Antimicrobial activity		CO_2 extraction	*S. typhimurium* LT2 strain MS1868 and *S. newington*.	1.25, 0.125, 0.0125, or 0.00125 μg/mL	Antibacterial activity of CBD against pathogenic bacteria.	Gildea et al., 2022

(Continued)

TABLE 15.1 CONTINUED
Medicinal Properties of *C. sativa*

Medicinal Property	Plant Part Used	Extraction Method	Model Used	Dosage Used	Effect Observed	Reference
Antimicrobial activity	Flower	Acetone and methanol extraction	Clinical strains of *S. aureus, P. aeruginosa, C. albicans,* and *A. niger*		Antimicrobial activity	Berardo et al., 2024
Antimicrobial activity		Ethanolic extract	*S. pseudintermedius* isolates and *S. pseudintermedius* ATCC 49051.	1–2 µg/mL	Reduced the number of viable bacterial cells by 99.9% at 6 and 24 h, respectively	Aiemsaard et al., 2022
Antimicrobial activity	Hemp inflorescence	Methanolic and aqueous extracts	Gram-positive and gram-negative bacteria and yeast-like fungus		Showed antibacterial activity	Motiejauskaitė et al., 2023
Anti-inflammatory			Human gingival fibroblast (HGF-1) cells	2.5 to 200 µg/mL and 1 mg/mL	Inhibited TNF-α and IL-1β production and promoted wound healing	Kongkadee et al., 2022
Anti-inflammatory	Flower	Ethanol exctraction	Rat models		Inhibition of paw oedema in rat models	Shebaby et al., 2021
Anti-inflammatory	Stem	Ethanol exctraction	Mice		Reduced TNF-α production and normalised activations of proapoptotic proteins in the liver, kidney, and spleen	Kim et al., 2022
Anti-inflammatory	Flower	Ethanol exctraction	Epithelial cells and colon cells	50 µL	Anti-inflammatory activity was found only for the low concentration of CBD	Nallathambi et al., 2017

(Continued)

TABLE 15.1 CONTINUED
Medicinal Properties of *C. sativa*

Medicinal Property	Plant Part Used	Extraction Method	Model Used	Dosage Used	Effect Observed	Reference
Heptoprotective	Seeds	Petroleum ether	CCl_4-induced	1 and 0.5 mL/kg	Showed scattered or slight areas of necrosis of the hepatocytes and slight fatty change	Ali et al., 2019
Heptoprotective	Leaves	Aqueous extract	Albino rats	10 and 20 mg/ kg/day	Normalized liver biochemical markers and minimised the hepatotoxic effects	Uzma et al., 2019
Heptoprotective	Hemp seeds	Soxhlet extraction	Female Wistar rats	10%	Alleviated fatty liver disease	Kaushal et al., 2020
Heptoprotective	Leaves	Hexane, DCM, and ethanol	Male albino rats	250 µg/mL	Showed protection against iron-induced oxidative liver damage	Erukainure et al., 2021
Cardiovascular properties	Female flowers	CO_2 extraction	Human	40 mg	CBD reduced phosphorylation of JNK, NF-κB, p70S6K, and STAT5, and increased phosphorylation of CREB, ERK1/2, Akt, and endothelial nitric oxide synthase (NOS)	Kicman, A. et al., 2020
Cardiovascular properties	Flowers	Ethanol	Human	320 µg/kg	↔ HR	Belgrave, B.E. et al., 1979
Cardiovascular properties	Root		Rat	60 nmol	↔ MBP, HR ↑ reflex bradycardiac response to BP increase (effect is dependent on 5-HT_{1A}) ↔ reflex tachycardiac response to BP decrease	Alves, F.H. et al., 2010

(Continued)

TABLE 15.1 CONTINUED
Medicinal Properties of *C. sativa*

Medicinal Property	Plant Part Used	Extraction Method	Model Used	Dosage Used	Effect Observed	Reference
Cardiovascular properties	Stem		Human isolated mesenteric arteries	10 μmol/L	Vasodilation (effect is independent on PPAR-γ)	Stanley, C.P. et al., 2014
Cardiovascular properties		Ethanol	Rat isolated perfused heart	30 μmol/L	↓ HR, contractility; arrythmias and asystole	Smiley, K.A. et al., 1976
Cardiovascular properties	Flowers	Ethanol	Human cardiomyocytes exposed to high glucose	4 μmol/L; 48 h	↓ oxidative and nitrative stress - ↓ apoptosis (mediated via modulation of Akt activity) - ↓ NF-κB activation	Rajesh, M. et al., 2010
Cardiovascular properties	Flowers		Mouse experimental autoimmune myocarditis	10 mg/kg; for 46 days; i.p.	↓ cardiac dysfunction ↓ myocardial fibrosis ↓ oxidative and nitrative stress in heart ↓ inflammation in heart ↓ mononuclear cell infiltration in heart ↓ necrosis in heart	Lee, W.S. et al., 2016

(*Continued*)

TABLE 15.1 CONTINUED
Medicinal Properties of *C. sativa*

Medicinal Property	Plant Part Used	Extraction Method	Model Used	Dosage Used	Effect Observed	Reference
Cardiovascular properties	Stem		Rabbit	0.1 mg/kg; 10 min	↓ blood troponin I ↓ dysfunction of left ventricle ↑ blood flow in the in the perfusion-defective region ↓ infarct size ↓ myocardial oedema and microvascular obstruction ↓ cardiac neutrophil infiltration ↓ apoptosis in heart	Feng, Y. et al., 2015
Cardiovascular properties			Piglet (newborn)	0.1 mg/kg; 15 min	↓ cerebral haemodynamic and metabolic impairment ↑ brain activity (EEG amplitude) ↓ seizures ↓ loss of neurons and neuron degeneration in cortex and hippocampus ↓ blood troponin T ↓ HI-induced decrease in MBP and increase in HR ↓ HI-induced blood gases and respiratory abnormalities	Alvarez, F.J. et al., 2008
Neuroprotective effects			*Caenorhabditis elegans*		On body bends rate was decreased in the wild-type (N2) strain after the 1.25% THC-treated group.	Vanin, A.P., et al., 2022

(Continued)

TABLE 15.1 CONTINUED
Medicinal Properties of *C. sativa*

Medicinal Property	Plant Part Used	Extraction Method	Model Used	Dosage Used	Effect Observed	Reference
Neuroprotective effects			Rats	10 mg/kg	Reduction of nigrostriatal degeneration accompanied by a damping of the neuroinflammatory response and an improvement of motor performance.	Giuliano, Claudio et al., 2021
Neuroprotective effects	Leaves	Hydroethanolic extract	Wistar rats	250 and 750 mg/kg	Aided in the management and treatment of stress induced neurodegeneration	Oladimeji, Olalekan et al., 2023
Hypertension	Seed		Rats	hemp seed meal	Strong hypotensive effects aided in the prevention and treatment of hypertension.	Girgih, A.T. et al., 2014
Depression	Seed	Ethanol extract	*Drosophila melanogaster*		Alleviated depression-like symptoms by modulating the circadian rhythm-related behaviours, transcript levels of neurotransmitter receptors, and neurotransmitter levels	Ahn, Yejin et al., 2021
Depression	Flower		Rats	15, 30, and 60 mg/kg	CBD in the 30 mg/kg dosage had similar effects to imipramine	Nibuya M et al., 1995
Antioxidant	Dried flowers	Ethanol extract	Stripped linseed oil	6.00 mg/mL	Hemp inflorescences can be used as a source of natural antioxidants in vegetable oils and lipid products	Carolina Cantele et al., 2020

cannabinoid and terpene levels in four cannabis varieties and their antiproliferative effects on four different cancer cell lines (BJ-5ta, BT16, HCC1806, and IMR5) using the MTT assay, which led to an increase in cannabinoid levels, particularly Δ9-tetrahydrocannabinol, and also increased terpene content in some cultivars. The MTT assay indicated that irradiation altered the anticancer properties of three out of the four tested extracts (Kovalchuk et al., 2020).

Cannabinoids have demonstrated anticancer effects in various models by inhibiting the proliferation, migration, and invasion of cancer cells and reducing tumour angiogenesis. However, cannabis also contains other compounds, such as terpenes and flavonoids, which have shown anti-tumourigenic effects. Cannabis co-treatment with THC or CBD and carfilzomib has been shown to lower the dose needed to impact cancer cell viability and migration, potentially allowing for lower doses and reducing side effects for patients in anticancer therapies (Tomko et al., 2020). In a study, Li et al., (2022) analysed 25 cannabis extracts with high levels of delta-9-tetrahydrocannabinol (THC) using HCC1806 squamous cell carcinoma and found that the extracts varied in their effectiveness, inhibiting cancer cell growth by 66% to 92%. Compounds from the cannabis plant, such as cannabinoids and terpenes, have shown antitumour effects in various types of cancer, including colorectal cancer (CRC). Terpenes like β-caryophyllene, limonene, and myrcene reported potential antitumour effects on CRC by promoting apoptosis and inhibiting cell proliferation and angiogenesis (Silva et al., 2023).

Ovarian cancer (OC) is a highly fatal gynecological malignancy, often diagnosed at an advanced stage. *C. sativa*, which produces more than 150 phytocannabinoids, including Δ9-trans-tetrahydrocannabinol (THC) and cannabidiol (CBD), has shown anticancer activity in various types of cancer *in vitro* and *in vivo* (Koltai et al., 2022). Cannabinoids have shown potential in cancer treatment by reducing cancer cell growth and invasion in some models, including mouse models. However, combining THC with TMZ significantly reduced tumour growth in T98 G xenografts, suggesting a potential role for cannabinoids in enhancing chemotherapy. Increasing CB1 and CB2 receptor expression in cancer cells may improve their response to cannabinoids like THC and CBD, but caution is necessary because of the possibility of antagonistic effects and receptor tolerance with repeated exposure. Additionally, cannabinoids showed promise in alleviating the side effects of cancer treatments and potentially enhancing the effects of chemotherapy and radiotherapy (Kovalchuk et al., 2020).

15.5.2 Antimicrobial Activity

Ethanol and oil extracts of cannabis leaves potently inhibited gram-positive bacteria (*Bacillus subtilis, Bacillus pumilus, Staphylococcus aureus, Micrococcus flavus*) and gram-negative bacteria (*Proteus vulgaris* and *Bordetella bronchiseptica*) (Głodowska et al., 2016). CBD showed strong inhibitory effects against gram-positive bacteria, with minimal inhibitory concentrations between 1 and 2 μg/mL, and exhibited potential bactericidal activity. CBD and CBDA were effective against gram-positive bacteria such as *S. aureus* and *Staphylococcus epidermidis*. CBDA showed lower antimicrobial activity than CBD, probably due to differences in its ability to penetrate bacterial cell membranes (Martinenghi et al., 2020). A study by Csakvari et al., (2021) synthesised silver nanoparticles (AgNPs) from cannabis, and the combination showed improved antioxidant properties and antibacterial activity against pathogens such as *Escherichia coli*, *Klebsiella pneumoniae*, *Pseudomonas fluorescens*, and *Staphylococcus aureus*.

Ethanolic and petroleum ether extracts of *C. sativa* leaves were effective against both gram-positive and gram-negative bacteria and fungi like *Candida albicans* and *Aspergillus niger*. Extracts from the leaves and stems of *C. sativa* had significant antibacterial effects against *Staphylococcus aureus*. The extracts could also inhibit the growth of various multidrug-resistant bacterial strains and some fungal strains, including methicillin-resistant *S. aureus* (MRSA) clinical isolates. *C. sativa* seed extracts were found to inhibit pathogenic bacteria without significantly affecting beneficial probiotic bacteria and also reduced the cell viability and biofilm formation of S. aureus (Schofs et al., 2021). Additionally, cannabinoids inhibit viral entry and activation by downregulating host cells' ACE2 receptors and TMPRSS2 enzymes, offering potential therapeutic benefits against COVID-19 (Mahmud et al., 2021). Cannabinoid receptor 2 agonists, especially phytocannabinoids, have been shown to have antimicrobial properties and are associated with the induction of microbiome composition flux (Hong et al., 2022). The standardised extracts of the Chinese strain of *C. sativa* were reported to be a promising candidate for developing new antibacterial agents to treat *S. aureus* infections (Muscarà et al., 2021).

Extracts from five different varieties of *C. sativa* exhibited antimicrobial and antifungal activities against gram-positive and gram-negative bacteria, yeast, and phytopathogenic fungi. Additionally, cannabis extracts can work synergistically with chemical fungicides, reducing the necessary doses of these chemicals (Berardo et al., 2024). The Moroccan *C. sativa* essential oil (EO) showed antimicrobial properties against pathogenic microorganisms. The antimicrobial evaluation revealed that *C. sativa* EO exhibits notable activity against various microorganisms, with minimum inhibitory concentrations (MICs) ranging from 1.2 to 37.8 mg/mL. When combined with fluconazole and ciprofloxacin, the EO significantly reduced their MICs, enhancing the antibiotics' effectiveness. These findings suggest that *C. sativa* EO could serve as a natural antimicrobial agent and might be beneficial in combination therapies to reduce the use of conventional antibiotics in treating infections (Nafis et al., 2019).

15.5.3 Anti-inflammatory Activity

The anti-inflammatory properties of essential oils enriched with terpenoid-rich precursors from three different cannabis chemotypes indicated that all three essential oils displayed moderate anti-inflammatory effects in the mouse paw swelling model but were less effective than CBD (Gallily et al., 2018). Terpenes with anti-inflammatory and aromatic properties are widely used in food additives and cosmetic products (Nuutinen et al., 2018), suggesting that terpenes may enhance the benefits of existing cannabinoid-based therapies (Downer et al., 2020). The cannabis oil extract (COE) significantly inhibited TNF-α production in LPS-stimulated mouse monocytes and reduced COX-2 and i-NOS protein expression while blocking the phosphorylation of MAPKs, including ERK, JNK, and p38 MAPK. Furthermore, COE effectively reduced paw oedema and inflammation in acute and chronic rat models. These results indicate that COE has two potent anti-inflammatory effects *in vitro* and *in vivo*, supporting its traditional use in treating inflammatory diseases such as arthritis (Shebaby et al., 2021). CBD's role as an inverse CB2 receptor agonist contributes to its anti-inflammatory effects. This study explores the chemical properties, bioavailability, and mechanisms of action of THC and CBD, emphasising their impact on oxidative stress and inflammation (Kopustinskiene et al., 2022).

The study by Kongkadee et al., (2022) suggested that hemp extract and CBD significantly reduced the release of TNF-α and IL-1β in a dose-dependent manner, reaching up to a 91.05% and 50.78% reduction in TNF-α, and up to 78.03% and 85.87% reduction in IL-1β, respectively, suggesting potential therapeutic benefits for treating oral inflammation and ulcers. Studies show that cannabis users tend to have a lower body mass index (BMI) than non-users, suggesting the plant's potential to reduce inflammation and treat comorbidities related to obesity.

The anti-inflammatory, antioxidant, and neuroprotective properties of phytocannabinoids make them promising candidates promising for further research as a treatment option for inflammation in obesity and other related conditions (Cavalheiro et al., 2022). The essential oils significantly reduced ear swelling in the xylene model, suggesting the inhibition of neutrophil accumulation. In the carrageenan model, CSEO decreased paw volume, indicating reduced oedema formation and leukocyte migration. In the CFA model, CSEO lowered contralateral paw volume, restored body weight, and reduced C-reactive protein levels, all signs of anti-inflammatory effects. These findings suggest that CSEO has promising antiarthritic and anti-inflammatory properties, making it a valuable potential therapy for managing arthritis and inflammatory diseases (Kabdy et al., 2023). The ethanolic extract of seeds exhibited significant inhibition of heat- and hypotonia-induced erythrocyte haemolysis, highlighting their potential therapeutic benefits (Haddou et al., 2023).

15.5.4 Hepatoprotective Activity

Cannabis expressed hepatoprotective properties and its potential benefits in managing and treating various liver diseases that were previously difficult to manage conservatively (Goyal et al., 2018). CCl_4-induced hepatotoxicity (administered in paraffin oil at 0.2 ml/kg for ten days) was significantly reduced by the co-administration of *C. sativa* oil in rats by lowering serum levels of AST, ALT, ALP, and bilirubin compared to the control groups. Also, the liver appeared to have less histological damage than the control and silymarin-treated mice (Ali et al., 2019). *C. sativa* aqueous extract significantly decreased serum levels of ALT, AST, and ALP, indicating reduced liver damage compared to the diclofenac-only group. In addition, *C. sativa* helped to restore the levels of red blood cells, white blood cells, platelets, and hematocrit, which were significantly reduced due to the toxicity of diclofenac (Uzma et al., 2019).

Hemp seed lipid fractions in hypercholesterolemic rats attenuated the morphological, histopathological, and biochemical signs of liver disease by inhibiting oxidative stress and inflammatory markers such as Cox-2, hPGDS, mPGES, IL-4, TNF-α, and sEH (Kaushal et al., 2020). Liver injury was induced *ex vivo* by exposing liver tissue to Fe^{2+}, which reduced the levels of antioxidants such as glutathione, superoxide dismutase, and catalase, as well as triglycerides and HDL-C, while increasing harmful markers such as malondialdehyde, nitric oxide, cholesterol, LDL-C, and some enzyme activities. Treatment with hexane, dichloromethane, and ethanolic extracts of *C. sativa* leaves significantly restored these levels and activities to almost normal levels, except for HDL-C. These extracts also improve glucose utilisation in liver cells (Erukainure et al., 2021).

15.5.5 Neuroprotective Activity

CBD and THC are beneficial in treating Alzheimer's disease by increasing the solubility of Aβ42 amyloid and Tau aggregation (Patil et al., 2023). Treatment of the PC12 cells with cannabidiol (10^{-7}–10^{-4}M) prior to β-amyloid peptide exposure significantly elevated cell survival while it decreased ROS production, lipid peroxidation, caspase-3 levels, DNA fragmentation, and intracellular calcium. Cannabidiol exerts a combination of neuroprotective, antioxidative, and anti-apoptotic effects against β-amyloid peptide toxicity, and the inhibition of caspase-3 appearance from its inactive precursor, pro-caspase-3, by cannabidiol is involved in the signalling pathway for this neuroprotection (Iuvone et al., 2004).

15.5.6 Hypertension

This work determined the ability of hemp seed meal protein hydrolysate (HMH)-containing diets to attenuate elevated blood pressure (hypertension) development in spontaneously hypertensive rats (SHRs). Effects of diets on plasma levels of renin and angiotensin I-converting enzyme (ACE) in the SHRs were also determined. Defatted hemp seed protein meal was hydrolysed using simulated gastrointestinal tract digestion with pepsin followed by pancreatin, and the resulting HMH was used as a source of antihypertensive peptides. The HMH was substituted for casein at 0.5 and 1.0% levels and fed to young growing rats for eight weeks (preventive phase) or adult rats for four weeks (treatment phase). Feeding young growing SHRs with HMH resulted in the attenuation of the expected increases in systolic blood pressure (SBP) with an average value of *120 mmHg when compared to the casein-only group of rats (control) with a maximum of 158 mmHg ($P < 0.05$). Feeding adult rats (SBP *145 mmHg) with identical diets for four weeks led to a significant ($P < 0.05$) reduction in SBP to *119 mmHg compared with 150 mmHg for the control rats. Plasma ACE activity was significantly ($P < 0.05$) suppressed (0.047–0.059 U/mL) in HMH-fed rats when compared to control rats (0.123 U/mL). Plasma renin level was also decreased for HMH-fed rats (0.040–0.054 lg/mL) compared to control rats fed only with casein (0.151 lg/mL). The results suggest that HMH with strong hypotensive effects in SHRs could be used as a therapeutic agent for both the prevention and treatment of hypertension (Girgih, A.T. et al., 2014).

15.5.7 Depression

In a study on depressed rats with a forced swim test, they stayed in the water without moving, unlike normal rats, which usually try to find a place to stay (de Mello Schier et al., 2014). Rats treated with CBD showed that CBD at a 30 mg/kg dosage had similar effects to imipramine, with a significantly higher number of rats climbing compared to rats treated with a saline solution and slightly higher numbers of rats than those that received other dosages of CBD. Additionally, a decrease in brain-derived neurotrophic factor (BDNF), a biomarker, was used to detect depression in animals and humans (Nibuya et al., 1995). Conversely, administration of antidepressant treatments increases BDNF expression, and brain infusion of BDNF produces antidepressant-like actions in rats (McArthur et al., 2006). Compatible with an antidepressant effect, an increase in BDNF was detected in animals treated with CBD and in animals treated with imipramine. Another study on an animal model (El-Alfy et al., 2010) was performed using FST and found the same antidepressant-like effect in a dose-dependent manner and a non-hallucinogenic effect in rats, unlike the rats that received tetrahydrocannabinol (THC).

15.5.8 Antioxidant Activity

Antioxidants protect the body from the effects of free radicals, inhibit the oxidation of biomolecules, and prevent cell damage (Tripathy et al., 2017; Nafis et al., 2019). The antioxidative effect of *C. sativa* extract on malondialdehyde and glutathione concentrations, as well as catalase activity in BALB/c mice, was monitored, wherein the extract significantly decreased the concentration of GSH in the blood of the mice and the concentration of MDA in the brain and liver of the mice compared to the respective concentrations in the control group. It also significantly increased CAT activity in the brain and liver of the mice after the metal-induced prooxidant effect, indicating that *C. sativa* is a good source of natural antioxidants and can be used in managing oxidative stress (Kubiliene et al., 2021).

15.5.9 Cardiovascular Activity

Cannabidiol (CBD) is a non-intoxicating and generally well-tolerated constituent of Cannabis that exhibits potential beneficial properties in a wide range of diseases, including cardiovascular disorders. Due to its complex mechanism of action, CBD may affect the cardiovascular system in different ways (Kicman et al., 2020).

15.5.9.1 Effects of Cannabidiol on the Cardiovascular System under Physiological Conditions

The administration of phyto-, endo-, and synthetic cannabinoids has a diverse and sometimes polyphasic influence on blood pressure (BP) and heart rate (HR) depending on the species, route of administration, presence of anaesthesia, and other experimental conditions. Cannabinoids can affect cardiovascular function not only via cannabinoid receptors but also through a variety of other receptors in the nervous system and directly in the blood vessels and heart. Stimulation of central CB1 receptors causes an increase in blood pressure, whereas peripheral CB1 receptors located presynaptically at the endings of pre- and/or postganglionic sympathetic neurons innervating the heart and vascular resistance are responsible for the hypotensive effects of cannabinoids. Activation of CB1 receptors located in the myocardium reduces contractility. In addition, cannabinoids can stimulate or inhibit the Bezold-Jarisch reflex, characterised by short and strong bradycardia and hypotension, via TRPV1 and 5-HT3 receptors located on sensory vagal nerve fibres (Malinowska et al., 2012). Cannabinoids, in most cases, cause vasodilation in isolated blood vessels or perfused vascular beds, although vasoconstriction is also observed. These effects result from direct activation of CB1, TRPV1, PPARs, and the putative endothelial cannabinoid receptors. In addition, the role of CB2, GPR55, and 5-HT1A receptors in the vasoactive effects of cannabinoids has been revealed in some studies (Richter et al., 2018).

Cannabinoids can also activate orphan GPR18 localised peripherally in blood vessels and centrally in the rostral ventrolateral medulla (RVLM), resulting in vasorelaxation and hypotension (Sierra et al., 2018). Cannabinoids may also affect the cardiovascular system through their metabolites, e.g., prostanoids. Thus, vasorelaxation or vasoconstriction evoked by cannabinoids was mediated indirectly through prostaglandin E receptor 4 (EP4) and prostacyclin receptor (IP) or prostaglandin E receptor 1 (EP1) and thromboxane receptor (TP), respectively (Stanley et al., 2014).

15.5.9.2 Effects of Cannabidiol on the Cardiovascular System under Pathological Conditions

The endocannabinoid system does not seem significant for cardiovascular regulation under physiological conditions, as both FAAH inhibitors and CB1 receptor antagonists do not significantly affect blood pressure in normotensive animals. This situation changes in pathological conditions when activation of the endocannabinoid system is often observed. Such activation may be protective or detrimental; for example, endocannabinoid-induced vasorelaxation is beneficial in arterial hypertension but deleterious in septic shock or portal hypertension. Pathological states might also modify the action of administered cannabinoids, for instance, the depressor response to THC is more potent in hypertensive than in normotensive patients (Malinowska et al., 2019). Moreover, the influence of cannabinoids on the cardiovascular system may be achieved through their modulating effect on immune processes or redox balance occurring via cannabinoid and non-cannabinoid receptors. Cannabinoids can cause oxidative stress and proinflammatory effects (mainly through CB1 receptors) as well as antioxidative and anti-inflammatory effects (mainly through CB2 receptors). Cannabidiol is known to possess antioxidant and anti-inflammatory properties. Thus, it might have therapeutic potential in treating different cardiovascular diseases, as oxidative stress and inflammation are essential to their pathogenesis (Atalay et al., 2019).

15.5.9.3 Effects of Abnormal-Cannabidiol on the Cardiovascular System

Abnormal-cannabidiol is a synthetic regioisomer of cannabidiol that did not exhibit overt behavioural effects in preliminary studies. Abn-CBD produces hypotension in anaesthetised dogs and mice after systemic administration and vasodilation in isolated mouse and rat mesenteric beds, rat mesenteric arteries, and rat, rabbit, and human pulmonary arteries. These effects are not dependent on cannabinoid receptors. The vasorelaxant action of Abn-CBD occurs, hypothetically, via an unidentified endothelial receptor called Abn-CBD receptor or endothelial cannabinoid receptor and is inhibited by CBD and O-1918, the putative antagonists of this receptor. However, the existence of a specific endothelial receptor for Abn-CBD is somewhat controversial. Abn-CBD is an agonist of GPR18, and O-1918 and CBD serve as antagonists or partial agonists/antagonists of this receptor. Therefore, some authors suggest that the proposed endothelial cannabinoid receptor may be GPR18. However, not all experimental observations confirm it explicitly. In addition, GPR18-independent activation of high-conductance Ca2+-activated K+ (BKCa) channels might contribute to the vasodilatory action of Abn-CBD. Abn-CBD can also lower blood pressure after intra-RVLM administration via GPR18 activation, which leads to sympathoinhibition. Moreover, this compound is an agonist of GPR55; however, this receptor does not mediate the vasorelaxant response to Abn-CBD (Kicman et al., 2020).

15.6 ADVERSE EFFECTS

The use of cannabis exposes the user to several compounds, unlike drinking or smoking, and it is currently classified as a Schedule I controlled substance by the FDA and used for restricted medical purposes to kerb abuse. Regardless of this, it is one of the most used substances among teenagers and adolescents in the United States. It is used as a means of social assimilation, for experimentation, enjoyment, and to cope with stress-related anxiety, depression, and post-traumatic experiences. About 4% of the global population was reported to use Cannabis in 2015 (Patel and Marwaha, 2024).

THC and CBD are the two critical psychoactive compounds in cannabis. They exert their effects by interacting with CB1 and CB2 receptors in humans, which are more prevalent in neurons and immune cells. Apart from inhibiting neurotransmitters like gamma-aminobutyric acid (GABA) and glutamate, these compounds modulate the neuronal system involving the reward, memory, learning, and pain pathways (Urits et al., 2020). Studies have shown that exposure to THC as early as adolescence can result in neurophysiological deficits by diminishing neuronal growth factors, affecting synapse formation, and loss of grey matter (Dhein, 2020). The risk of addiction towards cannabis increases with age and frequency of usage (Dhein, 2020). Globally, 13 million cannabis users meet the criteria for cannabis use disorder (CUD) caused by recurrent use of cannabis despite having adverse consequences, resulting in clinically significant impairments or distress (Thomsen et al., 2022). Habitual use of cannabinoids, natural or synthetic, results in a range of physiological issues, including sleep disorders, (hypo)manic symptoms, compulsive behaviour, depression, anxiety, hostility, or psychoticism. Apart from this, adolescent users are found to be associated with socioeconomic consequences, including lower work commitment and low income during their adulthood (Dhein, 2020). Persistent use of cannabis during adolescence risks induction of schizophrenia in adulthood, and genetically predisposed individuals are at higher risk. Several studies revealed that the carriers of the DRD2 rs1076560 T allele, functional COMTVal-158 polymorphism, COMTMet/Met, and AKTC/C genotype are likely to have higher chances of psychotic disorder when exposed to THC (Volkow et al., 2016). A study by Lin et al. revealed that prolonged usage of cannabis also had a relative risk of incidence of atrial fibrillation similar to that of a traditional tobacco user (Lin et al., 2022).

15.7 CONCLUSION AND FUTURE PERSPECTIVE

Cannabis is one of the oldest and most widely used psychoactive substances worldwide and has been legalised for adult use in countries like Canada, Uruguay, Mexico, and selected states in the United States, as well as for medical purposes in a few other countries (Global Cannabis Report, 2021). On the downside, cannabis use has been known to increase the risk of substance addiction and psychotic disorders. Interestingly, studies have also revealed its potential antioxidant effect, antibacterial properties, proinflammatory activity, and dose-dependent antidepressant effect, as well as its role in regulating the cardiovascular system and combating arterial hypertension. Extensive use of cannabis results in poor mental health and socioeconomic consequences. Studies have also shown that exposure to cannabis in patients with a predisposed risk of mental health issues may induce psychotic problems. There is a need for regulated availability of cannabis as teenagers and youth are more easily influenced. Hence, it is imperative to spread awareness in public to weigh the pros and cons of long-term usage of cannabis in medicinal and recreational use.

REFERENCES

Aiello A, Pizzolongo F, Scognamiglio G, Romano A, Masi P, Romano R. Effects of supercritical and liquid carbon dioxide extraction on hemp (Cannabis sativa L.) seed oil. *Int J Food Sci Technol*. 2020, 55, 2472–2480.

Aiemsaard J, Singh R, Borlace GN, Sripanidkulchai B, Tabboon P, Thongkham E. Antibacterial activity of cannabis extract (Cannabis sativa L. subsp. indica (Lam.)) against canine skin infection bacterium Staphylococcus pseudintermedius. *ScienceAsia*. 2022, 48, 348–353.

Ali SA, Sharief NH, Mohamed YS. Hepatoprotective activity of some medicinal plants in Sudan. *Evidence-Based Complementary and Alternative Medicine*. 2019, 2019, 2196315.

Amaducci S, Zatta A, Raffanini M, Venturi G. Characterisation of hemp (Cannabis sativa L.) roots under different growing conditions. *Plant Soil*. 2008, 313, 227–235.

Anwar F, Latif S, Ashraf M. Analytical characterization of hemp (Cannabis sativa) seed oil from different agro-ecological zones of Pakistan. *J Am Oil Chem Soc*. 2006, 83, 323–329.

Atalay S, Jarocka-Karpowicz I, Skrzydlewska E. Antioxidative and anti-inflammatory properties of cannabidiol. *Antioxidants* 2019, 9, 21.
Babiker EE, Uslu N, Al Juhaimi F, Ahmed IAM, Ghafoor K, Özcan MM, Almusallam IA. Effect of roasting on antioxidative properties, polyphenol profile and fatty acids composition of hemp (Cannabis sativa L.) seeds. *LWT.* 2021, 139, 110537.
Berardo ME, Mendieta JR, Villamonte MD, Colman SL, Nercessian D. Antifungal and antibacterial activities of Cannabis sativa L. resins. *J Ethnopharmacol.* 2024, 318, 116839.
Bouarfa M, Lebtar S, Boukhira S, Bousta D. An ethnobotanical and ethnopharmacological survey of Cannabis sativa of Taounate Region in Northern Morocco. *Int J Pharm Sci Rev Res.* 2020, 64, 116–122.
Callaway JC. Hempseed as a nutritional resource: An overview. *Euphytica.* 2004, 140, 65–72.
Campiglia E, Radicetti E, Mancinelli R. Plant density and nitrogen fertilization affect agronomic performance of industrial hemp (Cannabis sativa L.) in Mediterranean environment. *Ind Crops Prod.* 2017, 100, 246–254.
Carus M, Sarmento L. The European Hemp Industry: Cultivation, processing and applications for fibres, shivs, seeds and flowers. *Eur Ind Hemp Assoc.* 2016, 5, 1–9.
Cavalheiro EK, Costa AB, Salla DH, Silva MR, Mendes TF, Silva LE, Turatti CD, Bitencourt RM, Rezin GT. Cannabis sativa as a treatment for obesity: From anti-inflammatory indirect support to a promising metabolic re-establishment target. *Cannabis and Cannabinoid Research.* 2022, 7, 135–151.
Chakraborty S, Afaq N, Singh N, Majumdar S. Antimicrobial activity of Cannabis sativa, Thuja orientalis and Psidium guajava leaf extracts against methicillin-resistant Staphylococcus aureus. *Journal of Integrative Medicine.* 2018, 16, 350–357.
Clarke RC, Merlin MD, Small E. Evolution and classification of Cannabis sativa (Marijuana, Hemp) in relation to human utilization. *Bot Rev.* 2015, 81, 189–294.
Csakvari AC, Moisa C, Radu DG, Olariu LM, Lupitu AI, Panda AO, Pop G, Chambre D, Socoliuc V, Copolovici L, Copolovici DM. Green synthesis, characterization, and antibacterial properties of silver nanoparticles obtained by using diverse varieties of Cannabis sativa leaf extracts. *Molecules.* 2021, 26, 4041.
Da Porto C, Decorti D, Tubaro F. Fatty acid composition and oxidation stability of hemp (Cannabis sativa L.) seed oil extracted by supercritical carbon dioxide. *Ind Crops Prod.* 2012, 36, 401–404.
de Mello Schier AR, de Oliveira Ribeiro PN, Coutinho SD, Machado S, Arias-Carrión O, A Crippa J, W Zuardi A, E Nardi A, C Silva A. 2014. Antidepressant-like and anxiolytic-like effects of cannabidiol: A chemical compound of Cannabis sativa. *CNS & Neurological Disorders-Drug Targets (Formerly Current Drug Targets-CNS & Neurological Disorders).* 2014, 13, 953–960.
Dhein S. Different effects of cannabis abuse on adolescent and adult brain. *Pharmacology.* 2020, 105, 609–617.
Downer EJ. Anti-inflammatory potential of terpenes present in Cannabis sativa L. *ACS Chem Neurosci.* 2020, 11, 659–662.
El Khomsi M, Dandani Y, Chaachouay N, Hmouni D. Ethnobotanical study of plants used for medicinal, cosmetic, and food purposes in the region of Moulay Yacoub, Northeast of Morocco. *J Pharm Pharmacogn Res.* 2022, 10, 13–29.
El-Alfy AT, Ivey K, Robinson K, Ahmed S, Radwan M, Slade D, Khan I, ElSohly M, Ross S. Antidepressant-like effect of Delta(9)-tetrahydrocannabinol and other cannabinoids isolated from Cannabis sativa L. *Pharmacol Biochem Behav.* 2010, 95, 434–442.
Erukainure OL, Matsabisa MG, Salau VF, Oyedemi SO, Oyenihi OR, Ibeji CU, Islam MS. Cannabis sativa L.(var. Indica) exhibits hepatoprotective effects by modulating hepatic lipid profile and mitigating gluconeogenesis and cholinergic dysfunction in oxidative hepatic injury. *Frontiers in Pharmacology.* 2021, 12, 705402.

Flores-Sanchez IJ, Verpoorte R. Secondary metabolism in cannabis. *Phytochem Rev.* 2008, 7, 615–639.

Gallily R, Yekhtin Z, Hanuš LO. The anti-inflammatory properties of terpenoids from cannabis. *Cannabis and Cannabinoid Research.* 2018, 3, 282–290.

Gildea L, Ayariga JA, Ajayi OS, Xu J, Villafane R, Samuel-Foo M. Cannabis sativa CBD extract shows promising antibacterial activity against Salmonella typhimurium and S. newington. *Molecules.* 2022, 27, 2669.

Girgih AT, Alashi A, He R, Malomo S, Aluko RE. Preventive and treatment effects of a hemp seed (Cannabis sativa L.) meal protein hydrolysate against high blood pressure in spontaneously hypertensive rats. *Eur J Nutr.* 2014, 53, 1237–1246.

Global Cannabis Report (2021). https://f.hubspotusercontent10.net/hubfs/3324860/Reports/NFD-GlobalCannabisReport.pdf

Głodowska M, Łyszcz M. Cannabis sativa L. and its antimicrobial properties–A review. Conference: Research and Development of Young Scientists in Poland - Agronomy and Plant Protection At: Lublin. 2016, 77–82.

Goyal H, Rahman MR, Perisetti A, Shah N, Chhabra R. Cannabis in liver disorders: A friend or a foe?. *European Journal of Gastroenterology & Hepatology.* 2018, 30, 1283–1290.

Haddou S, Mounime K, Loukili E, Ou-Yahia D, Hbika A, Idrissi MY, Legssyer A, Lgaz H, Asehraou A, Touzani R, Hammouti B. Investigating the biological activities of Moroccan Cannabis sativa L seed extracts: Antimicrobial, anti-inflammatory, and antioxidant effects with molecular docking analysis. *Moroccan Journal of Chemistry.* 2023, 11, 11–14.

Hong H, Sloan L, Saxena D, Scott DA. The antimicrobial properties of cannabis and cannabis-derived compounds and relevance to CB2-targeted neurodegenerative therapeutics. *Biomedicines.* 2022, 10, 1959.

Hourfane S, Mechqoq H, Bekkali AY, Rocha JM, El Aouad N. A comprehensive review on Cannabis sativa ethnobotany, phytochemistry, molecular docking and biological activities. *Plants.* 2023, 12, 1245. https://doi.org/10.3390/plants12061245.

Irakli M, Tsaliki E, Kalivas A, Kleisiaris F, Sarrou E, Cook CM. Effect of genotype and growing year on the nutritional, phytochemical, and antioxidant properties of industrial hemp (Cannabis sativa L.) seeds. *Antioxidants.* 2019, 8, 491.

Iuvone T, Esposito G, Esposito R, Santamaria R, Di Rosa M, Izzo AA. Neuroprotective effect of cannabidiol, a non-psychoactive component from Cannabis sativa, on β -amyloid-induced toxicity in PC12 cells. *J Neurochem.* 2004, 89, 134–141.

Kabdy H, Azraida H, Agouram F, Oufquir S, Laadraoui J, Baslam A, Aitbaba A, Ouazzani ME, Elyazouli L, Aboufatima R, Garzoli S. Antiarthritic and anti-inflammatory properties of Cannabis sativa essential oil in an animal model. *Pharmaceuticals.* 2023, 17, 20.

Kaushal N, Gupta M, Kulshreshtha E. Hempseed (Cannabis sativa) lipid fractions alleviate high-fat diet-induced fatty liver disease through regulation of inflammation and oxidative stress. *Heliyon.* 2020, 6, e04422.

Kicman A, Toczek M. The effects of cannabidiol, a non-intoxicating compound of cannabis, on the cardiovascular system in health and disease. *Int J Mol Sci.* 2020, 21, 6740.

Kim YN, Sim KS, Park S, Sohn HY, Kim T, Kim JH. In vitro and in vivo anti-inflammatory effects of Cannabis sativa stem extract. *J Medicinal Food.* 2022, 25, 408–417.

Klauke AL, Racz I, Pradier B, Markert A, Zimmer AM, Gertsch J, Zimmer A. The cannabinoid CB2 receptor-selective phytocannabinoid beta-caryophyllene exerts analgesic effects in mouse models of inflammatory and neuropathic pain. *Eur Neuropsychopharmacol.* 2014, 24, 608–620.

Knapp AA, Lee DC, Borodovsky JT, Auty SG, Gabrielli J, Budney AJ. Emerging trends in cannabis administration among adolescent cannabis users. *J Adolesc Health.* 2019, 64, 487–493.

Koltai H, Shalev N. Anti-cancer activity of Cannabis sativa phytocannabinoids: Molecular mechanisms and potential in the fight against ovarian cancer and stem cells. *Cancers*. 2022, 14, 4299.

Kongkadee K, Wisuitiprot W, Ingkaninan K, Waranuch N. Anti-inflammation and gingival wound healing activities of Cannabis sativa L. subsp. sativa (hemp) extract and cannabidiol: An in vitro study. *Archives of Oral Biology*. 2022, 140, 105464.

Kopustinskiene DM, Masteikova R, Lazauskas R, Bernatoniene J. Cannabis sativa L. Bioactive compounds and their protective role in oxidative stress and inflammation. *Antioxidants*. 2022, 11, 660.

Kovalchuk O, Kovalchuk I. Cannabinoids as anticancer therapeutic agents. *Cell Cycle*. 2020, 19, 961–989.

Kovalchuk O, Li D, Rodriguez-Juarez R, Golubov A, Hudson D, Kovalchuk I. The effect of cannabis dry flower irradiation on the level of cannabinoids, terpenes and anti-cancer properties of the extracts. *Biocatalysis and Agricultural Biotechnology*. 2020, 29, 101736.

Kubiliene A, Mickute K, Baranauskaite J, Marksa M, Liekis A, Sadauskiene I. The effects of Cannabis sativa L. Extract on oxidative stress markers in vivo. *Life*. 2021, 11, 647.

Li D, Ilnytskyy Y, Ghasemi Gojani E, Kovalchuk O, Kovalchuk I. Analysis of anti-cancer and anti-inflammatory properties of 25 high-THC cannabis extracts. *Molecules*. 2022, 27, 6057.

Lin AL, Nah G, Tang JJ, Vittinghoff E, Dewland TA, Marcus GM. Cannabis, cocaine, methamphetamine, and opiates increase the risk of incident atrial fibrillation. *Eur Heart J*. 2022, 43(47), 4933–4942.

Liu FH, Hu HR, Du GH, Deng G, Yang Y. Ethnobotanical research on origin, cultivation, distribution and utilization of hemp (Cannabis sativa L.) in China. *Indian J Tradit Knowl*. 2017, 16, 235–242.

Lone TA, Lone RA. Extraction of cannabinoids from Cannabis sativa L. plant and its potential antimicrobial activity. *Univers J Med Dent*. 2012, 1, 51–55.

Lynch RC, Vergara D, Tittes S, White K, Schwartz C, Gibbs MJ, Ruthenburg TC, DeCesare K, Land DP, Kane NC. Genomic and chemical diversity in Cannabis. *Crit Rev Plant Sci*. 2016, 35, 349–363.

Mahmud MS, Hossain MS, Ahmed AF, Islam MZ, Sarker ME, Islam MR. Antimicrobial and antiviral (SARS-CoV-2) potential of cannabinoids and Cannabis sativa: A comprehensive review. *Molecules*. 2021, 26, 7216.

Malinowska B, Baranowska-Kuczko M, Schlicker E. Triphasic blood pressure responses to cannabinoids: Do we understand the mechanism? *Br J Pharmacol*. 2012, 165, 2073–2088.

Malinowska M, Toczek M, Pędzińska-Betiuk A Schlicker E. Cannabinoids in arterial, pulmonary and portal hypertension–Mechanisms of action and potential therapeutic significance. *Br J Pharmacol*. 2019, 176, 1395–1411.

Martinenghi LD, Jønsson R, Lund T, Jenssen H. Isolation, purification, and antimicrobial characterization of cannabidiolic acid and cannabidiol from Cannabis sativa L. *Biomolecules*. 2020, 10, 900.

Mawla F, Khatoon S, Rehana F, Jahan S, Shelley MMR, Hossain S, Haq WM, Rahman S, Debnath K, Rahmatullah M. Ethnomedicinal plants of folk medicinal practitioners in four villages of Natore and Rajshahi districts, Bangladesh. *Am Eur J Sustain Agric*. 2012, 6, 406–416.

McArthur R, Borsini F. Animal models of depression in drug discovery: A historical perspective. *Pharmacol Biochem Behav*. 2006, 84, 436–452.

McPartland JM, Hegman W, Long T. Cannabis in Asia: Its center of origin and early cultivation, based on a synthesis of subfossil pollen and archaeobotanical studies. *Veg Hist Archaeobot*. 2019, 28, 691–702.

Moccia S, Siano F, Russo GL, Volpe MG, La Cara F, Pacifico S, Piccolella S, Picariello G. Antiproliferative and antioxidant effect of polar hemp extracts (Cannabis sativa L., Fedora cv.) in human colorectal cell lines. *Int J Food Sci Nutr*. 2020, 71, 410–423.

Motiejauskaitė D, Ullah S, Kundrotaitė A, Žvirdauskienė R, Bakšinskaitė A, Barčauskaitė K. Isolation of biologically active compounds from Cannabis sativa L. inflorescences by using different extraction solvents and evaluation of antimicrobial activity. *Antioxidants*. 2023, 12, 998.

Muscarà C, Smeriglio A, Trombetta D, Mandalari G, La Camera E, Occhiuto C, Grassi G, Circosta C. Antioxidant and antimicrobial activity of two standardized extracts from a new Chinese accession of non-psychotropic Cannabis sativa L. *Phytotherapy Research*. 2021, 35, 1099–1112.

Nafis A, Kasrati A, Jamali CA, Mezrioui N, Setzer W, Abbad A, Hassani L. Antioxidant activity and evidence for synergism of Cannabis sativa (L.) essential oil with antimicrobial standards. *Industrial Crops and Products*. 2019, 137, 396–400.

Nallathambi R, Mazuz M, Ion A, Selvaraj G, Weininger S, Fridlender M, Nasser A, Sagee O, Kumari P, Nemichenizer D, Mendelovitz M. Anti-inflammatory activity in colon models is derived from δ9-tetrahydrocannabinolic acid that interacts with additional compounds in cannabis extracts. *Cannabis and Cannabinoid Research*. 2017, 2, 167–182.

Nibuya M, Morinobu S, Duman RS. Regulation of BDNF and trkB mRNA in rat brain by chronic electroconvulsive seizure and antidepressant drug treatments. *J Neurosci*. 1995, 15, 7539–7547.

Nuutinen T. Medicinal properties of terpenes found in Cannabis sativa and Humulus lupulus. *Eur J Med Chem*. 2018, 157, 198–228.

Patel J, Marwaha R. Cannabis use disorder. [Updated 2024 Mar 20]. In: StatPearls [Internet]. Treasure Island (FL): StatPearls Publishing; 2024 Jan-.

Patil N, Chandel V, Rana A, Jain M, Kaushik P. Investigation of Cannabis sativa phytochemicals as anti-Alzheimer's agents: An in silico study. *Plants*. 2023, 12, 510.

Pellati F, Borgonetti V, Brighenti V, Biagi M, Benvenuti S, Corsi L. Cannabis sativa L. and nonpsychoactive cannabinoids: Their chemistry and role against oxidative stress, inflammation, and cancer. *BioMed Res Int*. 2018, 2018, 1691428.

Richter JS, Quenardelle V, Rouyer O, Raul JS, Beaujeux R, Gény B, Wolff V. A systematic review of the complex effects of Cannabinoids on cerebral and peripheral circulation in animal models. *Front Physiol*. 2018, 9, 622.

Ross SA, ElSohly HN, ElKashoury EA, ElSohly MA. Fatty acids of cannabis seeds. *Phytochem Anal*. 1996, 7, 279–283.

Ryz NR, Remillard DJ, Russo EB. Cannabis roots: A traditional therapy with future potential for treating inflammation and pain. *Cannabis Cannabinoid Res*. 2017, 2, 210–216.

Schilling S, Melzer R, McCabe PF. Cannabis sativa. *Curr Biol*. 2020, 30, R8–R9.

Schofs L, Sparo MD, Sanchez Bruni SF. The antimicrobial effect behind Cannabis sativa. *Pharmacology Research & Perspectives*. 2021, 9, e00761.

Shebaby W, Saliba J, Faour WH, Ismail J, El Hage M, Daher CF, Taleb RI, Nehmeh B, Dagher C, Chrabieh E, Mroueh M. In vivo and in vitro anti-inflammatory activity evaluation of Lebanese Cannabis sativa L. ssp. indica (Lam.). *J Ethnopharmacol*. 2021, 270, 113743.

Sierra S, Luquin N, Navarro-Otano J. The endocannabinoid system in cardiovascular function: Novel insights and clinical implications. *Clin Auton Res*. 2018, 28, 35–52.

Silva-Reis R, Silva AM, Oliveira PA, Cardoso SM. Antitumor effects of Cannabis sativa bioactive compounds on colorectal carcinogenesis. *Biomolecules*. 2023, 13, 764.

Small E, Cronquist A. A practical and natural taxonomy for Cannabis. *Taxon*. 1976, 25, 405–435.

Stanley C, O'Sullivan SE. Vascular targets for cannabinoids: Animal and human studies. *Br J Pharmacol*. 2014, 171, 1361–1378.

Strzelczyk M, Lochynska M, Chudy M. Systematics and botanical characteristics of industrial hemp Cannabis sativa L. *J Nat Fibers*. 2021, 19, 5804–5826.

Taaifi Y, Benmoumen A, Belhaj K, Aazza S, Abid M, Azeroual E, Elamrani A, Mansouri F, Serghini Caid H. Seed composition of non-industrial hemp (Cannabis sativa L.) varieties from four regions in northern Morocco. *Int J Food Sci Technol*. 2021, 56, 5931–5947.

Thomsen KR, Thylstrup B, Kenyon EA, Lees R, Baandrup L, Ewing SW, Freeman TP. Cannabinoids for the treatment of cannabis use disorder: New avenues for reaching and helping youth?. *Neuroscience & Biobehavioral Reviews*. 2022, 132, 169–180.

Tomko AM, Whynot EG, Ellis LD, Dupré DJ. Anti-cancer potential of cannabinoids, terpenes, and flavonoids present in cannabis. *Cancers*. 2020, 12, 1985.

Tripathy S, Mohanty PK. Reactive oxygen species (ROS) are boon or bane. *Int J Pharm Sci Res*. 2017, 8, 1.

Urits I, Gress K, Charipova K, Li N, Berger AA, Cornett EM, Hasoon J, Kassem H, Kaye AD, Viswanath O. Cannabis use and its association with psychological disorders. *Psychopharmacology Bulletin*. 2020, 50, 56.

Uzma RA, Ambreen A. Evaluation of the heaptoprotectice effect of aqueous extract of Cannabis sativa (marijuana) against diclofenac sodium induced hepatotoxicity. *J Toxicol Pharmaceut Sci*. 2019, 3, 1–6.

Volkow ND, Swanson JM, Evins AE, DeLisi LE, Meier MH, Gonzalez R, Bloomfield MA, Curran HV, Baler R. Effects of cannabis use on human behavior, including cognition, motivation, and psychosis: A review. *JAMA Psychiatry*. 2016, 73, 292–297.

Index

3,4',5-Trihydroxystilbene, 54
7-Hydroxymitragynine, 118–119, 129
β-Cubebene, 166, 169
β-Pinene, 142, 164

A

Acetylcholinesterase (AChE), 26, 27, 79, 125–126, 151
Adipogenesis, 103
Adiposity, 103–104
Aeromonas hydrophilia, 120
Agarwood, 23–29
Akt signalling, 48–49, 60, 78, 149, 151, 168–171, 188, 190
Alkaloids, 110, 116–128, 139–150
Aloeswood, *see* Agarwood
Alzheimer's disease (AD), 3, 40, 47–50, 63–64, 77, 97, 101, 126, 151, 169, 195
AMPK pathway, 48–49, 61, 63, 103
Amyloid-beta, *see* Alzhiemers Disease
Analgesia, 29, 86, 114, 116–119, 125, 165, 184
Angiogenesis, 3, 48–49, 72, 76, 192
Anti-ageing, 63
Antibacterial effects, 77, 88, 116, 139, 141, 149, 172, 193
Anti-cancer, *see* cancer
Anti-carcinogenic activity, *see* Cancer
Antidesma thwaitesianum, 55, 57
Antidiabetic, *see* Diabetes Melitus
Antifungal, 2, 27, 29, 117, 139, 172, 193
Anti-heavy metal toxicity, 107
Anti-inflammatory, 4, 38, 45, 107, 115–119, 139, 141, 170–174, 193–194
Antimicrobial activities, *see* Antibacterial; Antifungal effects
Anti-microbial effects, 107, 144–145
Anti-obesity effects, 139
Antioxidant Activity; *see also* reactive oxygen species
  enzymes, 126, 151, 169
Antitussive, 116–117, 161
Antiviral, 36, 77
Anxiety, 126, 152, 198–199
AP-1, 4, 45, 122, 125, 149
Apoptosis, 35, 37, 40, 48–50, 76–78, 85–87, 129, 143, 149, 162, 168, 185
Aquilaria, *see* Agarwood
Atherosclerosis, 62–63, 87, 171–172
ATP1 protein, 120
Atrial fibrillation, 199
Autoimmune disease, 3, 168
Autophagy, 40, 60–65, 149, 151, 168
Ayurvedic medicine, 26, 83, 93–94, 141

B

Bacillus subtilis, 120, 192
BacoMind, 194
Bacopa monnieri, 93–98, 110
Bacopasaponin C, *see Bacopa monnieri*
Bacopaside II, *see Bacopa monnieri*
Bacoside A, *see Bacopa monnieri*
Bcl-2 protein, 37, 76, 87, 102, 126, 143, 162, 168
Beta-amyloid, *see* Alzheiemer's disease
beta-amyrin, 34, 39–41
beta-sitosterol, 34, 72
Biofilm formation, 28, 193
Black pepper, 139–153
Breast cancer, 37, 76, 87, 109, 128, 142, 149, 162, 168, 185

C

C. elegans, 62, 97, 108
Cannabinoids, *see Cannabis sativa*
Cannabis sativa, 182–199
Cannabis seeds, *see Cannabis sativa*
Carminative, 27, 141, 161
Caspase-3, 49–50, 76, 87–88, 102, 144, 149, 195
Catalase (CAT), 102, 126, 150–151, 195–196
Cell cycle arrest, 76, 109, 149
Cell proliferation, 5, 28, 35, 48–49, 60, 87, 109, 168, 192
Cell viability, 29, 168, 192, 193
Centella asiatica, 2, 5, 94
Cerebral ischaemia, 163, 165, 169–170
Chemoprevention, 76
Citrus hystrix, *see* Kaffir lime
Cleistocalyx nervosum var. paniala, 106–110
Clerodendrum petasites, 4, 44–50
Clinical studies, 1, 64, 93–94, 97, 98, 153
Cognitive
  enhancement, 98
  function, 104, 169
  impairment, 50
Cold Extraction Methods
  maceration, 12, 14–15
  percolation, 13–14
  Soxhlet Extraction, 13–15

supercritical CO_2 Extraction, 15
Curcuma comosa Roxb., 100–105
Curcuma longa, 3, 5–7, 110
Cyanidin 3-glucoside, 107–108
Cyclooxygenase -2 (COX-2), 45, 60, 102, 121–123, 127, 170, 194–195
Cyclin D1, 49, 87, 149
Cytochrome C, 76, 143
Cytochrome P450, 36, 129

D

Depression, 126, 147, 152, 191, 196, 198–199
Diabetes mellitus, 4, 78, 83, 163–169
Digestive disorders, 7
Drying methods for herbs, 12

E

Ellagic acid, 84–87, 107–109
Ellagitannins, 84–87
Endoplasmic reticulum (ER) stress, 40, 41, 129
Epigenetics, 61
ERK1/2 signalling, 41, 104, 149
Escherichia coli (*E. coli*), 120
Essential oils, 5, 141–142, 153, 173, 193–194
Estrogen receptors, *see* Oestrogen receptors
Expectorant, 27, 161

F

Fallopia japonica, 56
Fatty acids, 75, 141, 185
Flavones, 84
Flavonoids, 34, 57, 85, 106–109, 116, 145, 150, 173, 192
Flavonols, 84
Flu, 161, 173
Folk medicine, *see* Traditional Medicine
Free radicals, *see* Reactive oxygen species (ROS)

G

Gallic acid, 84–85, 87, 107–110, 185
Gene regulation, 127
Ginger, *see Zingiberofficinale Roscoe*
Gingerols, *see Zingiberofficinale Roscoe*
Ginkgo biloba, 110
Glandular trichomes, 18, 183
Glucose
lowering effects, 168
metabolism, 109
toxicity, 29, 169
Glutamate, 39–41, 49–50, 77–78, 97, 198
Glycosides, 29, 118, 150
Gynaecological disease, 100

H

Hair loss treatment, 161
Headache, 57, 161
Health drinks, 106
Hepatoprotective activity, 3–4, 146, 150, 199
Hepatotoxicity, 4, 39, 150, 194
HER2, *see* Breast Cancer
Herbal Extraction Methods
crude extracts, 11, 34, 104, 172
traditional, 16–17
HIV, 3, 77, 117
Hot Extraction Methods
decoction/boiling, 1, 16, 83, 126, 149
distillation methods
destructive distillation, 19
steam distillation, 18
hydro-distillation, 19
infusion, 16
reflux extraction, 17
HPLC, 20–21, 34, 96, 107
Huntington's disease (HD), 63, 77
Hypercholesterolemia, 87
Hyperglycaemia, *see* Diabetes melitus
Hypertension, 63–64, 147, 151–152, 161, 168, 173, 191, 195–197, 199
Hypoglycaemia, *see* Diabetes melitus

I

IL-6, 40–45, 121, 122, 126, 170–171, 176
Immunomodulation, 4, 107
Inflammation, 2, 11, 44, 64, 75, 78–79, 100, 110, 116, 121, 141–142, 150, 170–171, 182, 185, 194
Insecticide, 139
Insulin
resistance/sensitivity, *see* Diabetes melitus

J

Japanese knotweed, *see Fallopia japonica*

K

Kaffir lime, 3, 160–174
Keap-1/Nrf2 pathway, 125
Kratom, 114–130

L

Lactobacillus rhamnosus, 120
Leucorrhoea, 161
Linalool, 72, 162, 164, 168, 170, 172
Lipid metabolism, 101, 102–103, 139Liver
cancer, 46, 76

protection, 146, 150
Lupeol, 34, 39–40, 162, 167, 170
Lythraceae, 83

M

Makrut, *see* Kaffir lime
Memory
enhancing effects, 49, 93–98, 104
impairment, 64, 79, 198
Mental health, 128, 199
Metabolic dysfunction, 109
Methicillin-resistant *S. aureus* (MRSA), 88, 120, 172, 186, 193
MIC (Minimal Inhibitory Concentration), 120, 192
Mitragyna speciosa, *see* Kratom
Morris water maze, 93, 104
Morus alba, *see* Mulberry
Mulberry (*Morus alba*), 4, 54–64

N

Narcotic properties, 114, 184
Natural cannabinoids, *see* Cannabinoids
Neurodegenerative diseases, 41, 63, 77, 102, 109, 151
Neuroinflammation, 40, 77–78, 122, 126
Neuroplasticity, 124
Neuroprotection, 98, 109, 114, 125, 169, 195
NF-κB, 45, 48, 122, 149, 170, 171, 188
Nitric oxide synthase (NOS), 188
Nnaphthoquinones, *see Rhinacanthus Nasustus*

O

Oestrogen receptors, 61, 104
Opioid receptors, 118–119, 123, 127, 128
Ovarian hormone deficiency, 100
Oxidative phosphorylation, 120
Oxidative stress, *see* Reactive oxygen species (ROS)

P

Pain management, 125
Paradols, 72, 75
Parkinson's disease (PD), 49, 77, 102, 126
PARP, 144, 149
Perfumery, 139
Photoageing, 108
Phytoalexin, 54
PI3K-AKT-mTOR pathway, 60, 78
Piperine, *see* Black Pepper
Piper nigrum Linn, *see* Black Pepper
Pomegranate, 83–90
Probiotics, 120–121
Prostaglandin E2 (PGE2), 45, 62, 102
Prostate cancer, 64, 86, 144, 149
Pseudomonas aeruginosa, 120, 149
Punica granatum L., *see* Pomegranate
Punicalagin, *see* Pomegranate
Punica protopunica, *see* Pomegranate

Q

Quercetin, 118–120, 185
Quinones, *see Rhinacanthus nasutus*

R

Reactive oxygen species (ROS), 39, 49, 87, 102, 107–108, 123, 168
Red wine, 54
Resveratrol (RV), 54–65
Rhinacanthus nasutus, 2, 6, 13, 34–41
Rubiaceae family, *see* Kratom
Rutaceae family, *see* Kaffir lime

S

Salmonella typhi, 120, 149–150, 172
Shogaols, 72–73, 79
Sigma-1 receptor, 97
Silkworm, 55–56
Silymarin, 1, 194
Sitosterol, 34, 37–39, 72, 94, 162, 168–172
Skin cancer, 46
Skin disease, 3, 27, 163, 173
Sleep disorders, 199
Spatial memory, 104
SREBP-1, 149
Staphylococcus aureus, 149, 150, 172
Stigmasterol, 34, 39
Streptococcus pneumoniae, 120
Superoxide dismutase (SOD), 123, 126, 150–151, 195
Synthetic cannabinoids, *see Cannabis sativa*

T

Tannins, 84–85, 87, 145, 150Tau, *see* Alzheimer's disease
Terpenes, 34, 100, 142, 162–164, 173, 185, 192, 193
Terpinen-4-ol, 162, 164
Thai traditional medicine, 1–7, 50, 57, 83–89, 139–153, 182–199
THC, *see Cannabis sativa*
TNF-α, 28, 40, 45, 86, 109, 121–122, 126, 169–172, 187, 194–195
Tobacco, 199
Topoisomerase, 35–37, 143

Traditional herbal medicine, *see* Thai traditional medicine
Triterpenes, 34

U

Urolithins, 109
Uterine disorders, 100
UV damage, 4, 54, 108

V

Vascular dilation, 56
Volatile oil, 5, 15, 18, 19, 72, 139

W

Wnt/β-catenin signalling, 104
Wound healing, 2–4, 6, 89, 90, 164, 187

Z

Zingiberaceae, *see Zingiber officinale*
Zingiber officinale, 72–73

For Product Safety Concerns and Information please contact our EU representative GPSR@taylorandfrancis.com Taylor & Francis Verlag GmbH, Kaufingerstraße 24, 80331 München, Germany